MR. SONU KUMAR & SIMRAN PATHANIA

AI in Action: How Generative Models are Transforming Innovation

First published by Mr. Sonu Kumar & Simran Pathania 2024

Copyright © 2024 by Mr. Sonu Kumar & Simran Pathania

All rights reserved. No part of this publication may be reproduced, stored or transmitted in any form or by any means, electronic, mechanical, photocopying, recording, scanning, or otherwise without written permission from the publisher. It is illegal to copy this book, post it to a website, or distribute it by any other means without permission.

Mr. Sonu Kumar & Simran Pathania asserts the moral right to be identified as the author of this work.

Mr. Sonu Kumar & Simran Pathania has no responsibility for the persistence or accuracy of URLs for external or third-party Internet Websites referred to in this publication and does not guarantee that any content on such Websites is, or will remain, accurate or appropriate.

First edition

*This book was professionally typeset on Reedsy.
Find out more at reedsy.com*

Dedication
To the pioneers of artificial intelligence,
who dared to dream beyond the boundaries of possibility,
and to the innovators, thinkers, and creators,
whose relentless curiosity continues to shape our future.
This book is also dedicated to those
embracing the transformative power of AI,
with the hope that it inspires the next wave of groundbreaking ideas.

Contents

Preface	iii
Introduction	iv
Chapter 1: Understanding the Basics of Generative AI	1
Introduction to Generative Models	3
Key Concepts of Deep Learning	5
Convolutional Neural Networks (CNNs): Unlocking the Secrets of Images	6
Recurrent Neural Networks (RNNs): Mastering Sequential Data	6
Generative Adversarial Networks (GANs): The Art of Creative Competition	7
Transformer Networks: Mastering Long-Range Dependencies	8
Deep Learning: The Fuel for Generative AI's Creativity	8
Demystifying GANs	9
Ethical Considerations	11
Chapter 2: Generative AI in the World of Creativity	15
Storytelling and Literature	22
AI as a Storyteller: Exploring the Narrative Potential	22
The Impact on Literature: New Forms and Creative Collaborations	23
AI's Role in Authoring: A Catalyst for Innovation	23
The Challenges of Authenticity and Human Expression	23
The Future of Literature: Embracing the Synergy	24
Exploring the Ethical Dimensions	24
The Literary Landscape in the Age of AI: A Shift in Perspective	25
Fashion Design and AI	25
AI in Gaming and Virtual Worlds	28

Chapter 3: The Role of Generative AI in Business Innovation	30
Transforming Customer Experiences	33
Marketing and Generative AI	35
AI in Operations and Efficiency	41
Future Business Models	44
Chapter 4: Generative AI in Science and Medicine	47
Personalized Medicine and AI	50
AI and Genomics	54
Advanced Imaging and Diagnostics	56
AIs Role in Public Health	58
Chapter 5: Preparing for the Future with Generative AI	61
Innovative Leadership in the AI Era	68
AI and Societal Impact	70
Charting the Path Forward	73
References	77

Preface

We are living in an era where technology is not just reshaping industries—it is redefining what it means to innovate. At the forefront of this revolution stands artificial intelligence, and within its realm, generative models are heralding a new chapter of creativity and problem-solving.

This book is an exploration of how generative AI, with its ability to produce art, design, code, and even entire ecosystems of thought, is breaking down barriers and unlocking potential in unprecedented ways. It delves into the science and magic behind these models, offering insights into their applications across various fields and highlighting the opportunities they bring to individuals and businesses alike.

More importantly, this book is not just a tribute to the technology but a call to action. It is an invitation for readers to engage with the transformative power of generative AI, to see beyond the tools and embrace the mindset that drives innovation.

As we journey through the stories, technologies, and possibilities laid out in these pages, my hope is that you, too, will feel inspired to dream bigger, think deeper, and harness the infinite potential of AI. Whether you are a seasoned professional, a curious learner, or an entrepreneur with a vision, this book is for you.

Let us embark together on this exploration of how AI in action is shaping the world we live in—and the one we aspire to build.

Introduction

The rise of artificial intelligence has ushered in a transformative era—one where machines are not only automating tasks but also mimicking human creativity and innovation. At the core of this transformation lies generative AI, a subset of AI that has captured imaginations worldwide with its ability to generate text, images, music, designs, and even code.

What once seemed like science fiction is now a reality: AI models writing novels, designing skyscrapers, crafting marketing campaigns, and solving scientific puzzles. Generative models, such as GPT, Stable Diffusion, and others, have become powerful tools for creators, engineers, and businesses, enabling new ways to approach challenges and discover solutions.

But this is not just about the technology itself. This is about the change it brings to the way we think, work, and innovate. Generative AI is democratizing creativity, offering anyone with a vision the means to bring their ideas to life. It is challenging industries to rethink their processes, making way for efficiency, personalization, and scalability.

In this book, we will explore how generative models work, the science behind them, and the revolutionary applications they have unleashed across domains like healthcare, education, entertainment, and business. We will also delve into the ethical dilemmas and societal shifts brought about by these technologies, discussing questions around bias, intellectual property, and the future of work.

Through examples, case studies, and thought-provoking insights, this book aims to provide not only a technical understanding but also a vision for how generative AI can drive human progress. Whether you're a tech enthusiast, a professional navigating this new frontier, or simply someone curious about the AI-driven future, this book will serve as your guide to understanding and leveraging this incredible technology.

The world of generative AI is one of endless possibilities. It is a playground for the curious, a toolbox for the creators, and a catalyst for the ambitious. Together, let's uncover how this remarkable innovation is reshaping our world—and how you can be a part of it.

We are standing on the threshold of a technological renaissance—an era where artificial intelligence has evolved from a supporting tool to a driving force of innovation. Among the myriad advancements in AI, one phenomenon has captured global attention: generative AI. Unlike traditional AI systems designed to process data or automate tasks, generative models create. They compose poetry, design products, craft visual art, write software code, and simulate complex systems. They are the architects of a new age, reshaping how we think, work, and innovate.

This book, **"AI in Action: How Generative Models are Transforming Innovation,"** is an exploration of this transformative technology. At its core, it is a journey into the heart of generative AI—how it works, what it enables, and the profound implications it holds for society, creativity, and industry.

Generative AI is no longer confined to research labs or tech giants. It has become accessible to anyone with a device and an idea. Whether it's an artist seeking inspiration, a business optimizing its operations, or a scientist tackling unsolved problems, generative AI is opening doors once thought impenetrable. Tools like OpenAI's GPT models, DALL·E, and others have blurred the boundaries between human creativity and machine intelligence. What once required years of training or significant resources can now be accomplished with a few prompts, leveling the playing field for creators and innovators worldwide.

But how did we arrive at this point? The story of generative AI is one of relentless curiosity and bold experimentation. It is rooted in the evolution of neural networks, advances in computational power, and the marriage of deep learning with vast datasets. In the chapters that follow, we will uncover the mechanisms that power these models, from the intricate dance of transformers and attention mechanisms to the training processes that enable them to generate human-like outputs.

Yet, understanding the technology is only part of the picture. This book also seeks to answer the pressing questions:

- **What does this mean for the future of work and creativity?**
- **How can businesses and individuals leverage generative AI to innovate?**
- **What ethical considerations must we address as this technology becomes more pervasive?**

We will explore how industries are adapting to the generative AI wave, from healthcare professionals using AI to design personalized treatments to filmmakers creating realistic visual effects. We will examine case studies where generative AI has solved problems at scale, inspired new products, and even transformed how we teach and learn.

However, no innovation comes without challenges. As we embrace generative AI, we must grapple with its ethical dilemmas. Who owns the output of a generative model? How do we address the biases ingrained in the data that these models learn from? What safeguards can we implement to ensure that these technologies are used responsibly? These questions are as vital as the technology itself, and this book will explore them with the depth and nuance they deserve.

Generative AI represents more than a technological breakthrough; it symbolizes a shift in how we perceive the relationship between humans and machines. As we delve into the opportunities and challenges, this book aims to inspire readers not only to understand the technology but also to envision their role in this evolving landscape.

Whether you are a developer, a creative professional, a business leader, or a curious learner, this book is your guide to navigating the generative AI revolution. Together, we will uncover the potential of this extraordinary technology and explore how it is shaping the future.

Welcome to the world of generative AI—where imagination meets innovation and the possibilities are endless.

Chapter 1: Understanding the Basics of Generative AI

The story of artificial intelligence (AI) is a tapestry woven with threads of ambition, frustration, and ultimately, triumph. It began in the mid-20th century, fueled by the audacious dream of creating machines that could think like humans. The early pioneers, driven by a blend of scientific curiosity and a yearning to unlock the secrets of the human mind, laid the foundation for what we now call AI. The first glimmer of AI's potential appeared with the development of **rule-based systems** in the 1950s and 1960s. These systems, often referred to as "expert systems," were designed to perform specific tasks by following a set of predefined rules. Imagine a system programmed to diagnose medical conditions based on a patient's symptoms. By meticulously codifying medical knowledge into rules, these early AI systems could mimic the decision-making process of doctors, albeit in a limited and predefined way. However, the limitations of rule-based systems quickly became apparent. They were inflexible, unable to learn or adapt to new situations. Their success was contingent upon the completeness and accuracy of the programmed rules, making them brittle and prone to errors when faced with unexpected inputs. The pursuit of more versatile and intelligent AI led to the development of **machine learning** in the 1980s. This paradigm shift ushered in a new era where machines could learn from data, rather than relying on predefined rules. Imagine a machine trained on a vast dataset of handwritten letters, able to identify the correct character even if it encounters slightly different styles or variations. This ability to learn from examples enabled machines to tackle more complex problems, expanding the scope of AI's capabilities. Machine learning algorithms, powered by statistical methods, became increasingly sophisticated, leading to groundbreaking advances in various fields. From **image recognition** and **natural language processing** to **spam filtering** and **fraud detection** , AI began to permeate our lives, often unseen, yet quietly transforming the way we interacted with technology. However, the pursuit of truly intelligent machines continued. The advent of **deep learning** in the 2000s marked another critical turning point. Inspired by the structure of the human brain, **deep neural networks** , with their intricate layers of interconnected nodes, revolutionized the way we approached AI. These networks, trained on massive datasets, could learn complex patterns and relationships, surpassing the capabilities of traditional machine learning algorithms. This advancement in deep learning paved the way for **generative models** . These models, instead of simply recognizing patterns, could actually generate new data that resembled the patterns they had learned. Imagine a machine trained on thousands of photographs of cats. This generative model could then create entirely new cat images that look remarkably realistic, even though they never existed in the real world. This ability to generate novel data, often indistinguishable from real data, ushered in a new era of possibilities for AI, from creative pursuits to scientific breakthroughs. Generative models are not just a technological marvel; they represent a philosophical shift in our understanding of AI. They are not merely tools for mimicking human intelligence but rather powerful instruments for augmenting our own creative capabilities. They enable us to explore uncharted territories, from designing innovative products to composing mesmerizing music, all with the guidance of AI. The evolution of AI has been a journey of constant exploration, a testament to the relentless pursuit of creating machines that can think, learn, and even create. From the early days of rule-based systems to the sophisticated generative models of today, each step has pushed the boundaries of what's possible, shaping the world around us in profound ways. This journey is far from over. The future of AI, especially generative AI, holds immense promise, promising to reshape our lives in ways we can barely imagine. As we delve deeper into the world of generative models, we must always remember the ethical considerations that accompany such powerful technologies. Responsible development and thoughtful application are crucial for ensuring that AI benefits humanity, contributing to a future where technology and creativity harmoniously coexist.

Introduction to Generative Models

Imagine a world where machines can not only learn from data but also create new things—paintings, music, even medical treatments. This is the realm of generative AI, a branch of artificial intelligence that empowers machines to generate novel content, mimicking human creativity and ingenuity. Generative models, the heart of this transformative technology, stand apart from traditional AI models that primarily focus on analysis and prediction. While traditional AI excels at tasks like recognizing patterns in data, classifying objects, or predicting outcomes, generative models take a more proactive approach. They learn the underlying patterns and structures within data and then use this knowledge to generate new, original content. Think of it like a painter learning the nuances of brushstrokes, color theory, and composition. Instead of simply analyzing existing paintings, they use their understanding to create entirely new works of art. Generative models are like the artistic AI, capable of generating text, images, audio, code, and even scientific discoveries—all by learning from vast datasets and mimicking the creative processes of humans. But how do these models work? At the core of most generative models lies deep learning, a powerful technique that allows machines to learn from vast amounts of data through interconnected layers of artificial neural networks. These networks, inspired by the structure of the human brain, are designed to learn complex patterns and representations from data, much like our brains process information. One of the most notable and exciting advancements in generative AI is the rise of Generative

Adversarial Networks (GANs). These models, introduced in 2014, are a game- changer in the AI world. They are based on a fascinating concept: two neural networks, a generator and a discriminator, engage in a constant battle, pushing each other to improve. The generator's role is to create new data samples, trying to fool the discriminator, which is trained to distinguish real data from the generated ones. As the generator gets better at creating more realistic data, the discriminator learns to be more discerning. This competitive process, akin to a creative dance, leads to increasingly sophisticated and realistic output. Think of it like a counterfeiter trying to create perfect fake banknotes while a detective tries to spot the flaws. The counterfeiter keeps improving their techniques, while the detective sharpens their detection skills. This ongoing struggle pushes both players to new levels of expertise, resulting in incredibly realistic counterfeits or, in the case of generative models, stunningly realistic generated content. GANs have proven to be incredibly versatile, with applications ranging from creating realistic images and videos to composing music, generating text, and even designing new materials. The possibilities seem endless, as researchers continue to explore the potential of GANs and other generative models. But the power of generative AI comes with a set of ethical considerations that need careful attention. The potential for misuse, such as creating deepfakes or generating biased content, raises concerns about authenticity, misinformation, and potential harm. It is crucial to develop ethical frameworks and guidelines to ensure responsible development and use of these powerful technologies. Generative AI represents a paradigm shift in how we interact with technology. It is no longer just about analyzing and understanding data; it's about creating new things, pushing the boundaries of creativity, and transforming how we approach innovation. From art and music to scientific breakthroughs, generative AI is poised to revolutionize numerous fields, opening up a world of possibilities that we can only begin to imagine.

Key Concepts of Deep Learning

Deep learning, a powerful subfield of machine learning, stands as the driving force behind generative AI's remarkable abilities. It's like giving AI a brain—a complex, multi-layered network of interconnected neurons that can learn and process information much like our own brains do. At the core of deep learning lie **artificial neural networks**, intricately designed structures inspired by the biological nervous systems of living organisms. These networks consist of numerous interconnected nodes, or neurons, organized in layers. Each neuron receives input from other neurons and performs a simple calculation, passing the result to neurons in the next layer. By adjusting the strength of connections between neurons, the network learns to recognize patterns and extract meaningful information from data.

Think of a neural network as a sophisticated puzzle solver. Each layer of the network represents a step in the puzzle-solving process. As data flows through the network, each layer gradually refines the information, ultimately leading to a solution—whether it's generating a realistic image, composing a musical piece, or crafting a convincing dialogue.

Deep learning has revolutionized generative AI by enabling models to learn complex patterns and relationships within data, leading to a new era of creativity and innovation. Let's explore some key concepts of deep learning that are fundamental to understanding generative AI:

Convolutional Neural Networks (CNNs): Unlocking the Secrets of Images

Imagine a computer trying to understand a picture. It sees a grid of pixels, but can it grasp the essence of what it's looking at? This is where convolutional neural networks, also known as CNNs, shine. These networks specialize in processing visual data, mimicking the way the human brain analyzes images. They employ convolutional filters that move across the image, extracting features such as edges, shapes, and textures. These features are then processed by subsequent layers, allowing the network to learn and recognize complex visual patterns. CNNs are the backbone of many generative AI models that create realistic images, videos, and even 3D objects. Think of the art-generating AI models that can create breathtaking landscapes, portraits, or abstract masterpieces. They rely on CNNs to learn from vast datasets of images, enabling them to capture the nuances of light, shadow, and form.

Recurrent Neural Networks (RNNs): Mastering Sequential Data

Now, let's shift our focus to language. If you think about how we understand language, we process words in a sequential manner, considering the context of previous words. RNNs, much like CNNs, are designed to process sequential data, but instead of images, they work with time-based sequences like text, speech, or music. They have a special "memory" that allows them to retain information from previous steps in the sequence, enabling them to understand the context and dependencies between elements. RNNs are instrumental in powering generative AI models that can create compelling written content, translate languages, and even compose music. They can analyze the patterns and structures of language, allowing them to generate coherent text, translate languages accurately, and even create melodies and harmonies.

Generative Adversarial Networks (GANs): The Art of Creative Competition

GANs, as we discussed earlier, are a unique type of generative model that utilizes a competitive approach to learn and generate data. They consist of two neural networks—a generator and a discriminator—engaged in a constant game of cat and mouse. The generator attempts to create realistic samples that mimic the real data, while the discriminator tries to identify whether a sample is real or generated. This dynamic rivalry drives the learning process, as the generator constantly strives to fool the discriminator, becoming increasingly adept at generating realistic outputs. The discriminator, in turn, becomes better at detecting fake data, pushing the generator to improve its creations. GANs have revolutionized generative AI, allowing for the creation of highly realistic images, videos, and even audio. They have found applications in various fields, from generating photorealistic images of nonexistent people to producing stunning 3D models, opening up exciting possibilities for creative expression and content generation.

Transformer Networks: Mastering Long-Range Dependencies

While RNNs are good at processing sequential data, they struggle with long-range dependencies—the relationships between words that are far apart in a sentence. Imagine trying to understand a complex sentence like "The dog chased the cat, which then ran up a tree, causing the bird to fly away." RNNs have difficulty remembering the connection between "dog" and "bird" because they are separated by multiple words. Transformers, a groundbreaking architecture, emerged to address this challenge. They use a mechanism called "attention" to analyze relationships between words across the entire sentence, regardless of their position. This allows transformers to capture long-range dependencies and understand the subtle nuances of language. Transformers have quickly become the go-to architecture for various language-based tasks, including machine translation, text summarization, and question answering. They are also powering generative AI models that can generate creative and coherent text, write stories, and even translate languages with unparalleled fluency.

Deep Learning: The Fuel for Generative AI's Creativity

Deep learning is the engine that drives the creativity of generative AI. By allowing these models to learn complex patterns and relationships from vast amounts of data, deep learning enables them to generate new and exciting content that mimics human ingenuity. The advanced neural network architectures like CNNs, RNNs, GANs, and transformers empower generative AI to create realistic images, videos, audio, and text, pushing the boundaries of human imagination. As deep learning continues to evolve, we can expect even more innovative and powerful generative AI models to emerge, transforming how we interact with technology, create art, and solve problems in the future.

Demystifying GANs

Imagine a world where machines can not only analyze data but also create new ideas, art, and even scientific discoveries. This is the realm of generative AI, a revolutionary branch of artificial intelligence that empowers machines to go beyond simply learning from data and actively generate new, original content. At the heart of this transformative technology lies a fascinating concept: Generative Adversarial Networks (GANs).

GANs are a type of deep learning algorithm that operates on the principle of a game-like competition between two neural networks. These networks, often called the "generator" and the "discriminator," are locked in a constant struggle, driving each other to improve. The generator's task is to create new data that mimics the real data it has been trained on. Think of it as an aspiring artist trying to forge masterpieces that are indistinguishable from the works of renowned masters. The discriminator, on the other hand, acts as a discerning critic, tasked with evaluating the authenticity of the generator's creations. Its goal is to identify the fakes, distinguishing them from real data samples.

This adversarial relationship is the key to GANs' success. As the generator produces increasingly convincing fake data, the discriminator becomes more adept at identifying the subtle flaws. In response, the generator must constantly evolve its techniques, producing even more realistic creations to outwit its critic. This continuous back-and-forth, a game of cat and mouse, pushes both networks to reach new heights of sophistication. The generator, through this constant struggle, becomes increasingly proficient at generating data that is remarkably indistinguishable from reality.

To understand this concept more clearly, let's consider a practical example. Imagine training a GAN to generate realistic images of cats. The generator, in its initial stages, might produce blurry, misshapen images that bear little resemblance to the real thing. The discriminator, easily recognizing these imperfections, labels them as fake. This feedback loop forces the generator to adapt, learning from its mistakes and gradually refining its creations. As the generator improves, the discriminator becomes more astute, forcing the generator to further refine its techniques.

Eventually, through this iterative process, the generator becomes capable of producing highly realistic cat images that even humans might struggle to differentiate from real photographs. The power of GANs lies in their ability to create data that is not only realistic but also novel and innovative. This capability has revolutionized fields such as computer vision, natural language processing, and even medical imaging. In computer vision, GANs are used to generate high-quality synthetic images for training machine learning models. These synthetic images are invaluable in situations where real data is scarce or expensive to acquire. In natural language processing, GANs are used to generate realistic text, powering chatbots, creative writing applications, and even language translation tools. In the realm of medicine, GANs are proving their worth by generating realistic medical images, aiding in the development of diagnostic tools and improving the accuracy of medical diagnoses. Beyond these specific applications, GANs have sparked a revolution in creativity. Artists are using GANs to generate unique and surreal artwork, pushing the boundaries of artistic expression. Musicians are using GANs to compose novel and captivating musical pieces, exploring new musical styles and genres. The potential of GANs in these creative pursuits is vast, promising a future where machines can collaborate with human artists to create works of art that are both aesthetically pleasing and intellectually stimulating.

The applications of GANs extend to the world of business as well. Companies are using GANs to personalize customer experiences, generating recommendations based on individual preferences. They are also using GANs to improve product design and development, creating new products that are tailored to specific customer needs. The ability of GANs to generate realistic data sets has also proved invaluable in market research, allowing businesses to gain insights into consumer behavior and trends. The rise of generative AI, fueled by GANs, is transforming not only our world but also our understanding of what is possible. From creating stunning works of art to revolutionizing scientific discovery, these algorithms are poised to reshape our approach to creativity, problem-solving, and even our perception of reality.

But like any powerful tool, GANs come with ethical considerations. The ability to generate realistic content raises concerns about the potential for misuse, such as creating fake news or deepfakes to manipulate public opinion. It is crucial to develop ethical guidelines and frameworks for the responsible development and deployment of these technologies. As we continue to delve deeper into the world of generative AI, it is essential to remember that this technology is not merely a tool but a collaborative partner. The future of AI will be shaped by the way we choose to interact with these powerful algorithms, leveraging their capabilities while staying vigilant about potential risks.

Ethical Considerations

The ethical landscape surrounding generative AI is complex and evolving. While these technologies hold immense promise for innovation and progress, they also raise critical concerns that require careful consideration.

The Specter of Bias and Fairness: Generative AI models are trained on vast datasets, often reflecting the biases inherent in the real world. These biases can be amplified and perpetuated by AI systems, leading to unfair or discriminatory outcomes. For instance, an AI model trained on biased data might perpetuate stereotypes in art generation or create discriminatory hiring practices.

The Challenge of Transparency and Explainability:

Generative AI models can be highly complex, making it difficult to understand their decision-making processes. This lack of transparency raises concerns about accountability and potential misuse. If we cannot understand how a model reaches its conclusions, it becomes challenging to identify and address potential biases or errors.

The Dilemma of Intellectual Property: As AI models become more sophisticated in generating creative content, questions arise about ownership and copyright. Who owns the rights to a piece of art generated by an AI model? Is it the creator of the model, the person who provided the input, or a collective authorship? Resolving these legal complexities is crucial to fostering a fair and equitable creative ecosystem.

The Threat of Deepfakes and Misinformation: Generative AI enables the creation of highly realistic synthetic media, such as deepfakes – manipulated videos or audio recordings that can be used for malicious purposes. Deepfakes can be used to spread misinformation, damage reputations, or even incite violence. This poses a significant threat to the integrity of information and trust in online content.

The Potential for Job Displacement: The rise of generative AI raises concerns about job displacement. As AI models become capable of automating tasks traditionally performed by humans, certain jobs could become obsolete. This raises important questions about workforce retraining, social safety nets, and the future of work in a rapidly evolving technological landscape.

The Importance of Ethical Frameworks: Addressing these ethical challenges requires a multifaceted approach involving collaboration between researchers, developers, policymakers, and society as a whole. Establishing clear ethical frameworks, guidelines, and regulations for the development and deployment of generative AI is crucial. These frameworks should prioritize:

Data Fairness and Bias Mitigation: Ensuring that training datasets are diverse, representative, and free from biases. Developing techniques to detect and mitigate biases in model outputs.

Transparency and Explainability: Promoting the development of models that are interpretable and whose decision-making processes can be understood.

Intellectual Property Rights: Establishing clear guidelines for ownership and copyright in AI-generated content, ensuring fair compensation for creators and protecting intellectual property.

Safeguards Against Misuse: Developing tools and techniques to detect and prevent the creation and dissemination of deepfakes and other forms of synthetic media used for malicious purposes.

Social Impact and Labor Considerations: Addressing the potential for job displacement and promoting equitable access to AI-powered tools and opportunities.

Embracing Responsible Innovation: Generative AI presents a transformative opportunity for progress and innovation across various fields. However, it is crucial to approach this technology with responsibility and ethical awareness. By engaging in open dialogue, developing robust ethical frameworks, and fostering collaboration between different stakeholders, we can harness the power of generative AI for the betterment of society while mitigating its potential risks.

The Role of Education and Awareness: One of the most crucial steps towards responsible innovation is to raise awareness about the ethical implications of generative AI. Educating individuals about the potential benefits and risks associated with these technologies is essential. This education should target not just tech professionals but also policymakers, educators, and the general public. By fostering a deeper understanding of the ethical landscape surrounding generative AI, we can empower individuals to engage in informed discussions and make responsible decisions.

Collaboration and Open Dialogue: The development and deployment of generative AI should be guided by a collaborative spirit, involving researchers, developers, policymakers, and civil society. Open dialogue and exchange of ideas are essential to address ethical challenges and find solutions that benefit all stakeholders.

Promoting Inclusivity and Accessibility: It is crucial to ensure that the benefits of generative AI are widely accessible and inclusive. This means addressing issues of equity and accessibility, making sure that these technologies are available to people from diverse backgrounds and socioeconomic circumstances.

The Importance of Continuous Monitoring and

Evaluation: The ethical landscape surrounding generative AI is constantly evolving. Continuous monitoring and evaluation are essential to identify emerging ethical challenges and adapt our approaches accordingly. This ongoing dialogue and assessment will ensure that we are developing and using these technologies responsibly and ethically.

Embracing a Future of Responsible AI: The future of generative AI is promising, holding the potential to revolutionize creativity, innovation, and progress. However, realizing this potential requires a commitment to responsible innovation and ethical awareness. By embracing a collaborative spirit, establishing clear ethical frameworks, and fostering open dialogue, we can ensure that generative AI is used to benefit society and create a brighter future for all.

Chapter 2: Generative AI in the World of Creativity

The intersection of artificial intelligence (AI) and art is a fascinating realm where the boundaries of creativity are being redefined. Generative AI, with its ability to learn from vast datasets and generate new content, is revolutionizing the art world, sparking both excitement and debate. Traditionally, art has been considered a uniquely human endeavor, an expression of individual emotions, experiences, and perspectives. However, AI is challenging this notion by demonstrating its capacity to create artworks that are not only visually compelling but also thought-provoking and intellectually stimulating. One of the most prominent examples of AI in art creation is the use of Generative Adversarial Networks (GANs). GANs consist of two neural networks: a generator and a discriminator. The generator creates new data samples, mimicking the patterns it has learned from a training dataset, while the discriminator tries to distinguish between real data and generated data. Through this adversarial process, the generator learns to produce increasingly realistic and sophisticated outputs. The possibilities of GANs in art are boundless. Artists are using them to create realistic portraits, abstract paintings, and even entire landscapes. Some artists are exploring how GANs can be used to generate unique and unexpected variations on classic artworks, creating new interpretations of masterpieces. For instance, in 2018, an AI-generated portrait titled "Portrait of Edmond de Belamy" sold for over $432,000 at Christie's auction house. This landmark event brought AI art into the mainstream, highlighting its growing influence in the art world. However, the use of AI in art is not without controversy. Some argue that AI-generated art lacks the soul and emotional depth of human-made art, while others raise concerns about the potential for AI to devalue human artistry and creativity. A key concern is the issue of authorship and ownership. If an AI generates an artwork, who is considered the artist? Is it the programmer who created the AI model, the person who trained the AI, or the AI itself? These questions are complex and lack clear-cut answers. Another ethical consideration is the potential for AI to be used to create art that is deceptive or even harmful. For example, AI could be used to generate deepfakes, which are manipulated videos or images that can be used to spread misinformation or damage reputations.

Despite these concerns, AI art is here to stay. Its impact on the art world is undeniable, and it is likely to continue evolving and expanding in exciting new ways. Beyond visual art, AI is also finding applications in music, literature, and other creative fields. For instance, AI-powered music generators can create original compositions based on user-defined parameters, while AI algorithms can be used to analyze text and generate stories or poems. The rise of AI in art is not simply about replacing human artists but about exploring new forms of artistic expression and pushing the boundaries of creativity. AI can act as a powerful tool in the hands of artists, providing them with new ways to experiment, explore, and express themselves. AI can help artists overcome creative blocks, generate new ideas, and experiment with different styles. It can also assist in tedious tasks like color mixing or composition, allowing artists to focus on the more conceptual and artistic aspects of their work. The future of AI in art is bright, and it is likely to continue to evolve and transform the art world in profound ways. Artists, art collectors, and art enthusiasts alike will need to grapple with the ethical and philosophical questions surrounding AI art while embracing its potential to create new and exciting forms of artistic expression.

A Brief History of AI in Art

The use of AI in art has a rich history, with roots dating back to the early days of computer science.

- **The 1950s and 1960s:** The first computer-generated art was created in the 1950s, with simple programs used to generate abstract patterns and images.
- **The 1970s and 1980s:** Artists began experimenting with computer graphics and algorithms to create more complex and sophisticated art.
- **The 1990s and 2000s:** The development of the internet and the rise of personal computers made it easier for artists to create and share digital art.
- **The 2010s and beyond:** The advent of deep learning and Generative Adversarial Networks (GANs) revolutionized the field of AI art, leading to the creation of incredibly realistic and expressive art.

Different Approaches to AI Art Creation

AI art creation can be broadly categorized into a few different approaches:

- **Style Transfer:** This technique involves transferring the style of one image onto another. For example, a photograph can be transformed to look like a painting by Vincent van Gogh.
- **Generative Adversarial Networks (GANs):** As discussed earlier, GANs are a powerful tool for generating new and original artworks.
- **Deep Learning Models:** Other deep learning models, such as Variational Autoencoders (VAEs), are also being used to create art. VAEs learn to represent data in a lower- dimensional space, which can be used to generate new data points.
- **Machine Learning Algorithms:** Machine learning algorithms can be trained to identify patterns in data and then use those patterns to create art. For example, an algorithm could be trained on a dataset

of Impressionist paintings and then use that knowledge to create new Impressionist-style artworks.

AI in Art and the Human Touch

While AI can create remarkable artworks, it's important to note that it's not replacing the human element in art. AI is a tool that can be used to enhance and extend human creativity, not to replace it.

- **Collaboration:** Some artists are collaborating with AI systems to create art. Artists may provide input to the AI system, such as providing a specific prompt or style, and the AI system then generates art based on that input.
- **Artistic Expression:** AI can be a powerful tool for artists to express their emotions, ideas, and perspectives. By using AI, artists can explore new creative possibilities and challenge traditional art forms.

The Future of AI Art

The future of AI art is vast and exciting. As AI technology continues to evolve, we can expect to see even more sophisticated and innovative AI-generated art.

- **Immersive Art Experiences:** AI could be used to create immersive art experiences, such as virtual reality installations that allow viewers to step into the artwork and interact with it.
- **Personalized Art:** AI could be used to create personalized art, tailored to the individual preferences and tastes of the viewer.
- **AI-Powered Art Museums:** AI could be used to curate and manage art museums, using algorithms to identify patterns in art history and create new and engaging exhibits. The possibilities of AI in art are only just beginning to be explored. As AI technology continues to advance, we can expect to see an explosion of new and exciting AI art forms in the

years to come.

Crafting Music with Machines

Imagine a world where music creation is no longer confined to the human touch. Enter the realm of AI-powered music, a landscape where algorithms, neural networks, and generative models are orchestrating melodies, composing harmonies, and shaping the future of sound. This isn't just about replacing human composers; it's about expanding the creative possibilities, pushing the boundaries of musical expression, and collaborating with machines in a way that enhances the human experience. One of the most captivating applications of generative AI in music is the creation of entirely new compositions. Imagine a software program that can learn the intricate patterns of a specific genre, analyze the nuances of a composer's style, and then generate a piece that seamlessly blends originality with familiar elements. This is the power of AI-powered music composition. These algorithms are not simply replicating existing music; they are learning the underlying rules of musical structure, harmony, and melody. They can analyze vast databases of existing compositions, identifying patterns and relationships that may not even be consciously perceived by human musicians. This allows AI to create music that is both familiar and unexpected, drawing upon the history of music while venturing into uncharted sonic territories.

Several notable projects demonstrate the transformative power of AI in music creation.

AI Composers Making Waves:

AIVA (Artificial Intelligence Virtual Artist): This AI composer has gained recognition for its ability to generate classical music in various styles. AIVA can create original scores for films, video games, and other projects, blending classical elements with innovative sonic textures.

Amper Music: This platform uses AI to generate custom music tracks tailored to specific needs. Whether it's a background score for a video or a unique composition for a specific mood, Amper Music offers a personalized approach to music creation.

Jukebox (by OpenAI): This AI model can generate music in a variety of styles, including pop, rock, hip-hop, and even opera. Jukebox's ability to capture the nuances of different genres has made it a powerful tool for music exploration and experimentation.

Beyond creating entirely new compositions, AI is also revolutionizing the way we interact with existing music. It's now possible to use AI to analyze musical pieces, identify key elements, and even generate variations or remixes based on the original. This opens up a world of possibilities for reimagining classic works, exploring different interpretations, and expanding the lifespan of existing music.

AI as a Musical Companion:

Google's Magenta Project: This project explores the intersection of machine learning and music, developing tools and techniques for AI-powered music generation and analysis. Magenta has released several open-source tools that allow musicians to experiment with AI and incorporate its capabilities into their creative process.

AI-powered music analysis software: Several software programs are using AI to analyze music and provide insights into its structure, harmony, and melody. This information can be invaluable for musicians who are looking to improve their skills, develop new compositions, or simply gain a deeper understanding of the music they enjoy.

AI-powered music recommendation systems: Platforms like Spotify and Apple Music are using AI to personalize music recommendations based on user preferences. This creates a more tailored listening experience and opens up listeners to new artists and genres they might not have discovered otherwise. The impact of AI on the music industry is profound and far-reaching. It has the potential to democratize music creation, making it accessible to anyone with an internet connection. It's also transforming the way musicians collaborate, allowing them to work with AI tools to create music in ways that were previously unimaginable.

However, the rise of AI in music is not without its challenges. Some argue that AI-generated music lacks the soul and emotion of human-composed music. Others worry about the impact on the livelihoods of musicians as AI becomes increasingly sophisticated.

Challenges and Opportunities:

The debate about originality: There's a growing debate about the definition of originality in AI-generated music. Does AI truly create something new, or is it simply recombining elements from existing music? This is a complex philosophical question that will likely continue to be debated as AI becomes more sophisticated.

The impact on musicians: So me musicians fear that AI could eventually replace human composers and performers. However, it's more likely that AI will become a valuable tool for musicians, allowing them to focus on creative tasks while AI handles the technical aspects of music production.

The ethical considerations: As AI becomes more involved in music creation, there are ethical questions to consider. Who owns the copyright to AI-generated music? What are the responsibilities of developers when AI generates music that is offensive or harmful? These are crucial questions that need to be addressed as the field of AI-powered music evolves. Despite these challenges, the future of AI in music is bright. With continued advancements in AI research and development, we can expect to see even more innovative and transformative applications of AI in the world of sound.

AI and the Future of Music:

Personalized music experiences: AI will enable the creation of music that is perfectly tailored to individual preferences. Imagine a world where you can request a piece of music that perfectly reflects your mood, your taste, and even your personal experiences.

Interactive music experiences: AI can create music that responds to user input, allowing for real-time collaboration between humans and machines. This could lead to new forms of musical expression and interactive performance experiences.

Music for the Metaverse: As virtual reality and augmented reality become more mainstream, AI will play a critical role in creating immersive and interactive musical experiences in virtual worlds.

In the years to come, the lines between human and machine creativity will continue to blur. AI will not replace human musicians, but it will empower them, inspire them, and offer them new ways to express their artistry. The future of music is a collaboration between human and machine, a harmonious symphony of creativity and innovation.

Storytelling and Literature

The emergence of generative AI has sparked a fascinating dialogue about the future of storytelling and literature. While some argue that AI poses a threat to the unique creativity of human authors, others see it as a powerful new tool that can expand the possibilities of narrative and enhance the literary landscape. Let's explore this intriguing intersection of technology and art.

AI as a Storyteller: Exploring the Narrative Potential

Generative models, particularly those based on deep learning, have shown remarkable abilities in crafting narratives. These models can learn from vast amounts of text data, identifying patterns, themes, and stylistic elements that characterize different genres and writing styles. By processing and analyzing this data, AI can then generate text that mimics human-written narratives, creating stories that are both coherent and engaging.

One compelling example is the novel "The Day the Machines Stood Still," written by the AI system "The Bot" in collaboration with a human co-author. This project demonstrated the potential for AI to co-create narratives, generating storylines, characters, and dialogue that contributed significantly to the final product.

The Impact on Literature: New Forms and Creative Collaborations

The introduction of AI into the literary world has already begun to reshape the landscape. We are witnessing the emergence of new literary genres, experimental forms, and creative collaborations between humans and machines. AI- generated poems, short stories, and even entire novels are becoming increasingly prevalent, challenging traditional notions of authorship and sparking lively discussions about the future of literature.

AI's Role in Authoring: A Catalyst for Innovation

Generative AI can be a powerful tool for human authors, serving as a catalyst for innovation and creativity. It can help writers overcome writer's block, generate ideas, and experiment with different writing styles. Imagine an author working alongside an AI system that suggests plot twists, character developments, or evocative descriptions, thereby expanding the author's creative horizons. For example, the AI-powered writing assistant "Hemingway Editor" analyzes text for clarity, conciseness, and readability, helping authors refine their prose and create more engaging narratives. Similarly, tools like "Grammarly" utilize AI to identify grammatical errors and suggest improvements, enhancing the overall quality of writing.

The Challenges of Authenticity and Human Expression

While AI can generate coherent and engaging narratives, the question of authenticity and human expression remains a central debate. Some critics argue that AI-generated stories lack the depth, complexity, and emotional resonance that comes from human experience and imagination. They contend that AI may be adept at mimicking patterns and styles, but it cannot truly capture the essence of human creativity. This raises important questions about the role of AI in literary creation. Should AI be considered an author in its own right? How can we distinguish between stories generated by AI and those written by humans? These are complex questions that require careful consideration as AI technology continues to evolve.

The Future of Literature: Embracing the Synergy

The future of literature lies in embracing the synergy between human creativity and AI technology. Instead of viewing AI as a threat, we can see it as a powerful tool that can enhance and expand the possibilities of storytelling. Imagine a future where authors collaborate with AI to create immersive virtual realities, interactive narratives, or personalized literary experiences tailored to individual readers. This future holds exciting possibilities for both authors and readers alike. AI can help us break free from traditional storytelling constraints, explore new narrative forms, and engage with literature in innovative and engaging ways.

Exploring the Ethical Dimensions

The rise of AI in literature also raises important ethical considerations. How do we ensure that AI-generated narratives are not used to spread misinformation or harmful stereotypes? How can we address issues of copyright and authorship in an era where AI systems can create original works? It is essential to engage in open and transparent dialogue about the ethical implications of AI in literature. We need to develop guidelines and regulations that ensure that AI technology is used responsibly and that the integrity of the literary landscape is preserved.

The Literary Landscape in the Age of AI: A Shift in Perspective

The emergence of AI in literature necessitates a shift in perspective. We must move away from the notion that AI is a replacement for human creativity and embrace it as a complementary force that can enhance and expand the possibilities of storytelling. By embracing this synergy, we can usher in a new era of literary innovation and engage with narratives in ways that were previously unimaginable. AI's ability to process vast amounts of data, generate diverse creative outputs, and learn from human feedback opens up exciting avenues for authors and readers alike. It allows us to explore uncharted territories of storytelling, experiment with new forms, and engage with literature in ways that were previously unimaginable. As we navigate this evolving landscape, it is crucial to approach AI with a critical and thoughtful perspective, ensuring that it is used responsibly and ethically. By harnessing the power of AI, we can create a future where literature continues to flourish, inspiring, informing, and entertaining generations to come.

Fashion Design and AI

The world of fashion is undergoing a dramatic transformation thanks to the arrival of artificial intelligence (AI). This isn't just about automating tasks; AI is empowering designers to explore uncharted creative territories, pushing the boundaries of style and inspiring new trends. It's a fascinating blend of technology and artistry that's reshaping the industry from the way we design clothes to how we interact with fashion. One of the most visible applications of AI in fashion is in **generative design** . AI algorithms can analyze massive datasets of fashion imagery, identifying patterns, colors, and textures. This data can be used to create new designs that are aesthetically pleasing, innovative, and cater to specific consumer preferences. Imagine an AI-powered system that can generate thousands of unique dress designs, each with its own unique silhouette, fabric, and embellishments, all in a matter of minutes. This ability to generate variations on a theme can help designers quickly explore multiple possibilities and refine their ideas, saving time and resources.

AI-powered pattern making is another area where AI is revolutionizing fashion design. Traditionally, pattern making involved meticulous hand-drawn designs, which could be a time-consuming and complex process. AI tools can now streamline this process by generating intricate patterns automatically, based on sketches or 3D models. This means designers can focus on the creative aspects of fashion, leaving the technical aspects to AI.

Virtual fashion is becoming increasingly popular, and AI is playing a key role in this trend. AI algorithms can create realistic 3D models of clothing and accessories, allowing consumers to try on clothes virtually before purchasing them. This not only enhances the online shopping experience but also reduces the need for physical samples, contributing to sustainability in the industry.

The influence of AI extends beyond the design process; it is also being used to **predict fashion trends** . By analyzing social media data, search queries, and consumer behavior, AI algorithms can identify emerging trends and patterns. This valuable information can help fashion brands forecast demand, optimize their product lines, and stay ahead of the curve.

AI-powered personal stylists are emerging as a new way for consumers to discover and explore fashion. These AI systems can analyze a user's style preferences, body shape, and budget to recommend personalized outfits and accessories. This personalized approach helps consumers find items that suit their taste and needs, making shopping more efficient and enjoyable.

Beyond the design and retail aspects, AI is also impacting **fashion production** . Smart factories equipped with AI- powered robots can automate tasks like cutting, sewing, and printing, leading to increased efficiency and reduced production costs. This shift towards automation also opens up new possibilities for sustainable manufacturing practices, reducing waste and minimizing environmental impact. The integration of AI into fashion is not without its challenges. **Ethical considerations** are paramount, as concerns arise about potential job displacement and the impact on traditional craftsmanship. The potential for AI to perpetuate existing biases in fashion trends is also a critical issue that needs to be addressed. Despite these challenges, the potential of AI to revolutionize fashion is undeniable. By empowering designers, informing trends, and enhancing the consumer experience, AI is shaping the future of fashion, pushing boundaries and creating exciting new opportunities.

Here are some real-world examples of how AI is being used in fashion:

DressX is a platform that allows users to buy and wear digital clothes. Using AI algorithms, DressX generates realistic 3D models of clothing that can be worn in virtual reality, augmented reality, and social media.

The Fabricant is a fashion tech company that uses AI to create virtual clothes. Their work includes creating the first- ever virtual fashion item that was sold at a major auction.

Stitch Fix is a personal styling service that utilizes AI to recommend clothing and accessories based on a user's preferences and style.

Zalando is a leading European online retailer that uses AI to personalize product recommendations, predict trends, and optimize its supply chain.

Levi's has partnered with Google to develop AI-powered jeans that can adapt to a wearer's body shape and movement. **Adidas** has partnered with Carbon, a company that uses 3D printing technology, to create customized sneakers.

These examples demonstrate the diverse ways in which AI is being used to innovate in the fashion industry. From design and production to retail and marketing, AI is transforming every aspect of the fashion world.

Beyond the immediate applications, it is important to consider the long-term impact of AI on the fashion industry . AI has the potential to democratize fashion, making it more accessible to a wider range of consumers. AI-powered platforms could enable independent designers to reach global audiences and challenge established fashion houses. However, it is crucial to ensure that the development and application of AI in fashion are guided by ethical principles. We need to ensure that AI systems are developed and used in a way that is fair, transparent, and inclusive. **The future of fashion is a fascinating blend of technology and creativity** . AI is not here to replace human designers, but to empower them and help them push the boundaries of creativity. As AI continues to evolve, we can expect to see even more exciting applications of this transformative technology in the fashion industry.

AI in Gaming and Virtual Worlds

The realm of gaming and virtual worlds has always been a playground for pushing the boundaries of technology and imagination. Generative AI is now stepping into this arena, bringing a new dimension of realism, creativity, and immersive experiences. Imagine games where environments are not just pre-designed but dynamically generated, where characters react and evolve in unpredictable ways, and where stories unfold organically based on player choices. This is the promise of generative AI in gaming. One of the most significant ways AI is transforming gaming is in the creation of realistic and dynamic environments. Imagine a game where you are exploring a vast, open world. Instead of pre-defined landscapes, AI models can generate procedurally generated terrains, forests, and cities, each with its own unique character. This allows for endless replayability and eliminates the feeling of repetition that can plague open-world games. Developers can use AI to create complex ecosystems where flora and fauna interact naturally, responding to environmental changes and player actions in real time. Imagine encountering a creature that reacts differently depending on your approach, making each encounter a unique and engaging experience.

AI is also revolutionizing the creation of non-player characters (NPCs). Traditionally, NPCs have been limited in their interactions and often exhibit repetitive behaviors. Generative AI, particularly techniques like deep learning and natural language processing, allows developers to create more sophisticated NPCs with distinct personalities, motivations, and backstories. These characters can engage in natural conversations, react realistically to events, and even develop relationships with players over time. This dynamic interaction creates a more immersive and engaging gaming experience, blurring the lines between player and character. The use of generative AI extends beyond the visual and interactive aspects of gaming. AI can now assist in crafting compelling storylines and quests. By analyzing vast amounts of data, AI models can generate unique storylines based on player choices and preferences. Imagine a game where your actions directly influence the direction of the narrative, creating a truly personalized experience. AI can also assist in designing quests that are dynamically tailored to the player's progress and skill level, ensuring a constant sense of challenge and engagement. The potential for generative AI in virtual worlds is even more profound. Imagine VR and AR experiences where environments, objects, and even the avatars themselves are generated in real time, adapting to your interactions and movements. This level of immersion can revolutionize how we interact with virtual spaces, blurring the lines between reality and the digital world. Imagine being able to design and furnish your virtual home, create custom clothing for your avatar, or even collaborate on projects with other users in a shared virtual space. However, the integration of AI in gaming also brings ethical considerations. Questions arise about the potential for AI to be used for manipulative or deceptive purposes. How can we ensure that AI-driven games are fair and equitable, avoiding biases that might disadvantage certain players? Furthermore, the use of AI to generate highly realistic characters and environments raises concerns about the potential for blurring the lines between reality and simulation. It's crucial to consider the potential psychological impact of AI-powered gaming experiences and to ensure that they remain ethical and responsible.

The future of gaming and virtual worlds will be shaped by generative AI. From creating more realistic and dynamic environments to crafting immersive narratives and interactive experiences, AI is poised to revolutionize how we play, interact, and explore digital worlds. As AI technologies continue to advance, we can expect to see even more innovative and engaging applications in the gaming industry, pushing the boundaries of creativity and entertainment. The journey has just begun, and the possibilities are truly limitless.

Chapter 3: The Role of Generative AI in Business Innovation

The realm of product development is undergoing a paradigm shift, fueled by the transformative power of generative AI. Businesses across industries are embracing this technology to unlock unprecedented levels of innovation, efficiency, and customer satisfaction. This section delves into the fascinating ways generative AI is reimagining product design and development, from conceptualization to prototyping and beyond. Imagine a world where designing a new product is no longer confined to the limitations of human imagination. With generative AI, this dream is becoming a reality. AI models can analyze vast datasets of existing products, consumer preferences, and market trends to generate novel designs that push the boundaries of creativity. This ability to go beyond traditional design constraints is empowering businesses to create products that are not only innovative but also better suited to the evolving needs of their target audiences.

One of the key advantages of generative AI in product development lies in its ability to rapidly explore a multitude of design possibilities. Traditional design processes often involve a series of iterative refinements, which can be time- consuming and resource-intensive. Generative AI, on the other hand, can generate countless design variations within a short period, allowing designers to quickly experiment with different forms, materials, functionalities, and aesthetics. This rapid iteration enables businesses to explore a wider design space, leading to more innovative and competitive products.

Beyond aesthetics, generative AI is also transforming the core functionality of products. By analyzing data on user interactions, market dynamics, and technological advancements, AI models can suggest design improvements that enhance product performance, usability, and overall customer satisfaction. For example, AI can help optimize the ergonomics of a chair, suggest improvements to the user interface of a mobile app, or even create new product features based on user feedback. The impact of generative AI extends beyond the product itself. It is also revolutionizing the way businesses approach product prototyping and testing. Traditionally, creating physical prototypes was a costly and time-consuming process. Generative AI is enabling businesses to create virtual prototypes that can be iterated upon and tested digitally, significantly reducing development costs and lead times. This virtual prototyping process also allows for more efficient user testing, as prototypes can be shared with potential customers for feedback before any physical production takes place. The benefits of generative AI in product development are not limited to large corporations. Startups and small businesses are also leveraging this technology to gain a competitive edge. With access to powerful AI tools and platforms, even smaller companies can now create products that compete with those developed by larger organizations. This democratization of innovation is driving a wave of entrepreneurship and creativity across industries.

Here are some concrete examples of how businesses are leveraging generative AI for innovative product development:

1. **Automotive Industry:** Generative AI is being used to design new car models with improved aerodynamics, fuel efficiency, and safety features. AI models can analyze data from wind tunnels and crash tests to generate designs that optimize performance and minimize environmental impact.
2. **Consumer Electronics:** Generative AI is helping to create innovative

smartphone designs that incorporate advanced features, such as AI-powered cameras and personalized user interfaces. AI models can analyze user behavior and preferences to suggest design improvements that enhance the overall user experience.
3. **Fashion Industry:** Generative AI is being used to design new clothing styles and patterns that are more sustainabl and tailored to individual preferences. AI models can analyze data on fashion trends, materials, and manufacturing processes to create designs that are both innovative and environmentally conscious.
4. **Medical Devices:** Generative AI is helping to develop new medical devices with improved precision and functionality. AI models can analyze data from clinical trials and patient records to suggest design modifications that enhance the effectiveness and safety of medical devices.
5. **Industrial Design:** Generative AI is being used to create new designs for industrial products, such as furniture,appliances, and machinery. AI models can analyze data on product performance, user needs, and manufacturing constraints to generate designs that are both functional and aesthetically pleasing.

These examples demonstrate the transformative potential of generative AI in product development. By leveraging AI's ability to analyze vast datasets, generate creative solutions, and optimize design parameters, businesses can create products that are more innovative, efficient, and customer- centric than ever before. As AI technology continues to evolve, we can expect even more groundbreaking applications in product development, further blurring the lines between human creativity and artificial intelligence. However, it is important to acknowledge that generative AI is not a silver bullet. While it offers significant advantages, it's essential to use it responsibly and ethically. Businesses must ensure that AI-generated designs are aligned with their brand values, legal requirements, and ethical considerations. Moreover, it's crucial to strike a balance between AI- generated creativity and human oversight, allowing human designers to refine and validate the AI's outputs. The future of product development is intertwined with generative AI. As this technology continues to evolve and become more accessible, we can expect to see a surge in innovation across all industries. Businesses that embrace generative AI and integrate it strategically into their product development processes will be best positioned to capitalize on this transformative wave and thrive in the rapidly evolving landscape of the 21st century.

Transforming Customer Experiences

In the realm of business, generative AI is not just a technological marvel; it's a game-changer poised to redefine customer experiences. This transformation hinges on AI's ability to understand and cater to individual customer needs, providing personalized interactions that resonate deeply. Let's delve into how generative AI is revolutionizing customer service and personalization, creating a new era of customer engagement. Imagine a world where every interaction with a business feels tailored specifically to you. Generative AI is making this vision a reality. It's no longer about one-size-fits-all solutions; it's about leveraging AI's power to create truly individualized experiences.

One of the most impactful ways generative AI is enhancing customer service is through AI-powered chatbots. These intelligent assistants are not simply automated scripts; they're capable of learning and adapting to each customer's needs. They can understand natural language, answer questions with remarkable accuracy, and even anticipate customer requests. This dynamic engagement eliminates the frustration of navigating clunky IVR menus and waiting on hold. Beyond chatbots, generative AI is transforming how businesses personalize content and recommendations. Imagine stepping onto a website and being greeted with product suggestions tailored to your past purchases and preferences. This level of personalization isn't just about offering relevant products; it's about fostering a sense of understanding and value. By analyzing vast amounts of customer data, generative AI can predict what customers might need or want, making every interaction feel like a thoughtful suggestion from a knowledgeable friend. The benefits of generative AI in customer service extend beyond personalized interactions. AI-powered tools can analyze customer feedback in real-time, identifying trends and pain points. This insight empowers businesses to proactively address customer concerns, improve their products and services, and build stronger customer relationships. Consider the impact of AI on customer support. By analyzing past interactions, generative AI can create detailed knowledge bases that empower customer support agents to provide accurate and efficient assistance. This enables agents to resolve customer issues faster, ensuring a seamless and positive experience.

Generative AI's ability to understand and respond to emotions is another game-changer. By analyzing the tone and sentiment of customer interactions, AI can detect frustration or dissatisfaction, allowing businesses to intervene and provide support before the situation escalates. This proactive approach fosters customer trust and loyalty. The potential of generative AI in personalized marketing is equally transformative. Imagine receiving targeted advertising based on your browsing history and interests, but not in a generic, intrusive way. Generative AI can create personalized marketing messages that feel relevant and engaging, making customers more likely to engage with brands and products. This personalization extends beyond advertising. Generative AI can be used to create customized product recommendations, tailored content, and even personalized offers. This individualized approach builds a stronger connection with customers, making them feel valued and understood. Generative AI also empowers businesses to engage with customers on a deeper level. By analyzing customer data, businesses can understand their customer's preferences, behaviors, and needs. This information can be used to create targeted marketing campaigns, personalized product offerings, and even tailor the customer experience on websites and mobile apps.

However, the integration of generative AI into customer experiences is not without its challenges. Ethical considerations are paramount. Businesses must ensure that they use AI responsibly, respecting customer privacy and data security. Transparency is crucial; customers need to understand how their data is being used and have control over their privacy. Moreover, businesses need to be mindful of potential biases in AI models. AI algorithms can inherit biases present in the data they are trained on, which can lead to discriminatory or unfair outcomes. It's crucial to implement mechanisms to mitigate bias and ensure that AI-powered systems treat all customers fairly. The successful integration of generative AI into customer experiences requires a shift in mindset. It's not simply about deploying AI technology; it's about embracing a customer- centric approach that prioritizes personalization, empathy, and transparency. Businesses that successfully leverage generative AI will be able to forge stronger customer connections, build loyalty, and drive sustainable growth. In conclusion, generative AI is revolutionizing customer experiences, creating a future where interactions are personalized, intuitive, and tailored to individual needs. By harnessing the power of AI, businesses can build a world where customers feel valued, understood, and empowered. However, this transformative potential comes with ethical considerations that must be addressed thoughtfully and responsibly. By striking the right balance between innovation and ethical practices, generative AI can truly empower businesses to create customer experiences that are both personalized and meaningful.

Marketing and Generative AI

The world of marketing is undergoing a profound transformation, driven by the immense power of generative AI. Imagine a world where marketing campaigns are crafted not just based on data analysis but on the creative insights generated by intelligent algorithms. This is the reality that generative AI is bringing to the forefront, opening up new possibilities for marketers to connect with consumers in more personalized, engaging, and impactful ways. Generative AI models, with their ability to learn from vast amounts of data and generate creative content, are empowering marketers to go beyond traditional approaches. They are not merely analyzing data but are actively creating compelling narratives, crafting captivating visuals, and tailoring messages to resonate with individual consumers. This revolution in marketing is driven by a fundamental shift in how we approach customer understanding and communication.

AI-Powered Content Creation: Crafting Engaging Narratives

One of the most significant impacts of generative AI on marketing is its ability to generate high-quality content, including written copy, images, and videos. Gone are the days of spending hours brainstorming ideas or relying on generic stock images. AI-powered content creation tools can now produce captivating content tailored to specific audiences and marketing objectives.

Imagine a scenario where a marketer needs to create a compelling product description for a new line of sustainable clothing:

- **Traditional Approach:** The marketer would spend time researching customer demographics, analyzing competitor products, and crafting a compelling narrative that highlights the brand's values and product benefits.
- **AI-Powered Approach:** The marketer can input key information about the product, target audience, and desired brand messaging into an AI-powered content generation tool. The tool can then generate multiple versions of product descriptions, complete with persuasive language and imagery, tailored to different customer segments.

This not only saves time and resources but also allows marketers to experiment with different creative angles and identify the most impactful messaging for specific audience segments.

Beyond Words: Visual Storytelling with AI

Generative AI is also revolutionizing visual marketing. With the emergence of AI-powered image generation tools, marketers can now create stunning visuals for their campaigns without the need for professional designers. This opens up a world of possibilities for creating unique and engaging content, from social media graphics to website banners to product mockups.

Consider the challenge of a small business owner launching a new line of handcrafted jewelry:

- **Traditional Approach:** The owner would need to hire a graphic designer to create visuals for their online store, social media posts, and marketing materials. This could be a costly and time-consuming process.
- **AI-Powered Approach:** The owner can input keywords describing their jewelry designs, target audience, and desired style into an AI-powered image generation tool. The tool can then generate multiple visual concepts, showcasing the jewelry in different settings and with various artistic styles, allowing the owner to select the most impactful visuals for their campaign.

AI-powered image generation not only democratizes visual creativity but also allows marketers to create highly personalized visuals tailored to different segments of their target audience.

AI-Powered Personalization: Reaching the Right Audience

The true power of generative AI in marketing lies in its ability to personalize the customer experience. By leveraging AI insights and algorithms, marketers can create personalized content, offers, and experiences that resonate with individual consumers. This level of personalization is unprecedented and is transforming how brands connect with their customers.

Imagine a scenario where an online retailer is launching a new line of cosmetics:

- **Traditional Approach:** The retailer would send generic email campaigns and social media promotions to all their subscribers, hoping to capture the interest of some.
- **AI-Powered Approach:** The retailer can use AI-powered recommendation engines to analyze each customer's past purchase history, browsing behavior, and preferences. Based on this data, the AI can generate personalized recommendations for each customer, suggesting products, offers, and content that are most likely to resonate with their individual interests.

This level of personalization not only increases the effectiveness of marketing campaigns but also strengthens customer relationships by demonstrating a genuine understanding of their needs and preferences.

AI for Predictive Marketing: Anticipating Consumer Needs

Generative AI is not only about creating engaging content and delivering personalized experiences; it's also about anticipating consumer needs and trends. By analyzing vast amounts of data, AI can identify emerging trends and consumer behaviors, allowing marketers to stay ahead of the curve and develop campaigns that resonate with the evolving market.

Consider the challenge of a beverage company looking to launch a new product:

- **Traditional Approach:** The company would rely on market research surveys and focus groups to gather consumer feedback and identify potential trends. This process can be time-consuming and expensive.
- **AI-Powered Approach:** The company can leverage AI- powered data analytics to analyze social media trends, online search data, and consumer reviews related to beverages. This data can reveal emerging preferences, flavors, and packaging trends, allowing the company to develop a product that is more likely to resonate with the target audience. Generative AI can help businesses develop more informed product roadmaps, tailor their marketing strategies to emerging trends, and stay competitive in a dynamic market.

Challenges and Considerations in Generative AI Marketing

While generative AI presents exciting opportunities for marketers, it's important to acknowledge the challenges and considerations associated with its adoption. Here are some key factors to keep in mind:

- **Data Privacy and Security:** Ensuring data privacy and security is paramount when using AI-powered marketing tools. It's essential to comply with data protection regulations and employ robust security measures to protect customer information.
- **Bias and Fairness:** AI models are trained on vast amounts of data, which can reflect existing societal biases. It's crucial to address potential biases in AI-generated content and ensure that marketing campaigns are fair and inclusive.
- **Transparency and Explainability:** AI models can be complex and their decision-making processes may not be readily understandable. It's essential to strive for transparency and explainability in AI-powered marketing, allowing marketers to understand and control the outcomes.

- **Ethical Considerations:** As AI takes on more creative and strategic roles in marketing, it's crucial to consider the ethical implications of its use. It's important to ensure that AI-powered marketing campaigns are ethical, responsible, and aligned with societal values.

The Future of Marketing with Generative AI

The future of marketing is inextricably linked with the advancement of generative AI. As AI technologies continue to evolve, we can expect even more transformative innovations in how brands connect with consumers.

- **Hyper-Personalized Marketing:** AI will enable hyper- personalized marketing experiences, tailoring content, offers, and interactions to individual customer preferences and needs.
- **AI-Driven Customer Service:** Generative AI will play an increasingly significant role in customer service, providing personalized support, answering questions, and resolving issues efficiently.
- **Interactive Marketing Experiences:** AI will enable interactive marketing experiences that engage consumers in real-time, creating personalized journeys that foster brand loyalty.
- **Predictive Marketing Insights:** AI will provide marketers with predictive insights into consumer behavior, enabling them to anticipate trends, optimize campaigns, and stay ahead of the competition.

The possibilities are endless, and the future of marketing is bright with generative AI at the forefront. By embracing AI's power and navigating its challenges responsibly, marketers can unlock new levels of creativity, engagement, and impact, forging deeper connections with consumers and driving business success.

AI in Operations and Efficiency

The realm of business operations is undergoing a transformative shift, fueled by the power of artificial intelligence (AI). Generative AI, with its ability to create new solutions and optimize existing processes, is at the forefront of this revolution. From streamlining workflows to enhancing customer interactions, AI is poised to redefine the very essence of business efficiency. One of the most profound impacts of AI on business operations is in **automation** . AI-powered tools are adept at automating repetitive tasks, freeing up human employees to focus on more strategic and creative endeavors. Consider a scenario where a customer service team is inundated with routine inquiries about product availability or order status. AI-driven chatbots can handle these queries efficiently, providing accurate information and resolving issues quickly. This not only saves time for the customer service team but also elevates the overall customer experience. Beyond simple automation, AI is revolutionizing **process optimization** . By analyzing vast amounts of data, AI algorithms can identify bottlenecks, inefficiencies, and areas for improvement within existing processes. For instance, in a manufacturing setting, AI can analyze sensor data from machines to predict potential failures before they occur. This predictive maintenance can minimize downtime, optimize production schedules, and reduce overall costs. AI's analytical prowess is also transforming **supply chain management** . By analyzing data from various sources, such as sales forecasts, supplier performance, and market trends, AI can optimize inventory levels, identify potential disruptions, and suggest more efficient routes for transportation. This leads to reduced costs, improved delivery times, and a more responsive supply chain.

Data-driven decision-making is another area where AI excels. By processing large datasets, AI can provide insights and recommendations that are often beyond the capabilities of human analysts. This enables businesses to make more informed decisions, based on real-time data and predictive analytics. Imagine a scenario where an e-commerce company uses AI to analyze customer purchasing patterns and identify emerging trends. This information can be leveraged to personalize product recommendations, tailor marketing campaigns, and optimize pricing strategies.

Generative AI is further amplifying the impact of AI on business operations. By creating new content, designs, and solutions, generative models are empowering businesses to innovate in ways never before possible. Consider a company that uses generative AI to design new product prototypes based on customer feedback. This process can significantly speed up the design cycle and lead to more innovative products that better meet market demand.

AI in Operations and Efficiency: Case Studies

To better understand the practical implications of AI in business operations, let's delve into real-world examples:

Amazon: Amazon's sprawling e-commerce empire is built on a foundation of AI-driven automation and optimization. Their fulfillment centers leverage robots and AI algorithms to manage inventory, pack orders, and ship products with remarkable efficiency. AI also powers their recommendation engine, providing personalized product suggestions to customers, driving sales and enhancing their online experience.

Netflix: Netflix utilizes AI extensively to personalize content recommendations for its subscribers. By analyzing viewing history, ratings, and other user data, AI algorithms suggest movies and shows that are likely to be of interest. This personalized experience has become a cornerstone of Netflix's success, driving user engagement and subscriptions. **Uber:** The ride-hailing giant employs AI to optimize its ride- matching system, ensuring efficient pickups, timely deliveries, and reduced wait times for both drivers and passengers. AI algorithms also help Uber analyze traffic patterns, predict surge pricing, and ensure driver safety.

Beyond the Current Horizon: The Future of AI in Operations

The application of AI in business operations is continuously evolving, with new possibilities emerging every day. Here are some potential future trends:

Hyper-automation: As AI matures, we can expect to see a rise in hyper-automation, where AI automates entire business processes, not just individual tasks. This could lead to significant efficiency gains and completely reimagine the way work is done.

AI-powered decision-making: AI is likely to play an even more central role in decision-making, offering real-time insights and recommendations for complex business problems. This could empower businesses to make more informed decisions, anticipate market shifts, and stay ahead of the competition.

Augmented human intelligence: Instead of replacing humans, AI is increasingly being used to augment human intelligence. By providing valuable insights, AI can empower employees to make better decisions, solve problems more effectively, and ultimately become more productive.

The Importance of Ethical Considerations

As AI becomes more integrated into business operations, it's crucial to address ethical considerations. Issues such as data privacy, algorithmic bias, and job displacement need to be carefully considered and addressed. Transparent data practices, fair algorithms, and strategies for reskilling workforces are essential to ensure that AI is used responsibly and ethically.

Embrace the AI Revolution

The revolution in business operations fueled by generative AI is already underway. Businesses that embrace AI and leverage its potential are well-positioned to thrive in the rapidly changing world. By automating tasks, optimizing processes, and unlocking new insights, AI is empowering businesses to become more efficient, innovative, and customer-centric.

Future Business Models

The advent of generative AI marks a paradigm shift in the business landscape, ushering in a wave of new possibilities and prompting a reimagining of traditional business models. As these models continue to evolve and become more sophisticated, they are poised to reshape industries, disrupt established paradigms, and pave the way for previously unimaginable innovations. One of the most significant impacts of generative AI on future business models lies in the realm of **product development** . Imagine a world where products are not merely designed by human engineers but co-created by AI systems. Generative models can analyze vast datasets of customer preferences, market trends, and material properties to suggest novel product concepts, optimizing for factors like cost, functionality, and sustainability. This opens up avenues for hyper-personalized products tailored to individual needs and preferences, blurring the lines between mass production and bespoke creation. Consider the example of a clothing company. Instead of relying solely on human designers, they can leverage generative AI to create virtual prototypes based on real-time fashion trends, customer feedback, and even individual body measurements. This allows for rapid iteration and experimentation, leading to more tailored and innovative designs. Moreover, AI can analyze patterns in sales data to predict future trends, enabling businesses to proactively develop products that resonate with evolving consumer desires. Beyond product development, generative AI is poised to revolutionize **customer experiences** . In a world increasingly driven by personalized interactions, generative models can create customized experiences across every touchpoint. Chatbots powered by advanced language models can engage in natural conversations, providing 24/7 customer support, answering complex questions, and resolving issues with remarkable efficiency. Generative AI can also personalize marketing campaigns, recommending products and services based on individual browsing histories, purchase preferences, and even social media interactions. This level of personalization fosters a sense of individual attention and increases customer engagement, ultimately driving loyalty and repeat business.

Marketing and sales will also undergo a significant transformation driven by generative AI. The ability to analyze vast amounts of data allows for the creation of highly targeted marketing campaigns that resonate with specific customer segments. Generative models can craft compelling content, including ad copy, social media posts, and even personalized emails, ensuring that marketing messages land effectively with the intended audience. AI-powered tools can analyze the performance of marketing campaigns in real-time, allowing businesses to optimize their strategies for maximum impact. Furthermore, generative AI can revolutionize **operational efficiency** by automating repetitive tasks and optimizing complex processes. Imagine a manufacturing facility where robots powered by generative AI autonomously adjust production lines based on real-time data, ensuring optimal efficiency and minimizing downtime. AI-driven predictive maintenance can identify potential equipment failures before they occur, preventing costly disruptions and ensuring smooth operations. In logistics, AI algorithms can optimize delivery routes, reducing transportation costs and delivery times, enhancing customer satisfaction. Beyond individual industries, generative AI is poised to reshape the entire **business landscape** , leading to the emergence of new business models and industry structures. Imagine a world where companies don't just sell products but offer tailored experiences, combining AI-powered services with physical goods. This shift towards "experience-as-a-service" opens up a whole new realm of possibilities, where businesses can monetize their expertise and data rather than solely focusing on product sales. One potential example is the rise of "AI-powered design studios" that offer customized design solutions for businesses across industries. These studios leverage generative AI to rapidly prototype products, generate branding assets, and even create personalized marketing materials, empowering clients to bring their visions to life with unprecedented speed and efficiency.

The implications of generative AI extend far beyond the realm of business. It holds the potential to **democratize access to creativity and innovation** , empowering individuals to create and express themselves in new and exciting ways. Imagine a world where anyone can leverage AI to write a novel, compose a song, or design a virtual world, fostering a new era of artistic collaboration and expression.

This democratization has the power to **reshape the workforce**, requiring a shift towards skills that complement and collaborate with AI. As machines become increasingly adept at performing routine tasks, the demand for human skills like critical thinking, creativity, and emotional intelligence will grow. This creates opportunities for new roles that focus on strategic planning, creative problem- solving, and ethical AI governance.

However, the widespread adoption of generative AI also raises important **ethical considerations**. It is imperative to ensure that these technologies are developed and deployed responsibly, prioritizing fairness, accountability, and transparency. Addressing bias in AI algorithms, protecting intellectual property, and ensuring data privacy are crucial aspects of ethical AI development.

The future of business with generative AI is brimming with potential, promising a world where creativity, innovation, and personalized experiences are at the forefront. By embracing this transformative technology, businesses can unlock new opportunities, redefine industry landscapes, and usher in a new era of growth and prosperity.

Chapter 4: Generative AI in Science and Medicine

Imagine a world where the arduous task of developing life-saving drugs is no longer a years-long, painstaking process, but a journey guided by the sharp insights of artificial intelligence. This is the reality being shaped by the revolutionary power of generative AI in drug discovery, a field where innovation is desperately needed to address global health challenges. The traditional drug discovery pipeline is a complex and often inefficient process. It involves screening vast libraries of chemical compounds, testing their efficacy and safety in preclinical studies, and finally conducting clinical trials on human volunteers. This process can take years, costing billions of dollars and often failing to yield effective treatments for many diseases. Generative AI is poised to transform this landscape by injecting a dose of speed, precision, and intelligence into drug discovery. How? By utilizing machine learning algorithms that can analyze massive datasets of scientific information, identifying patterns and connections that might escape human observation. These algorithms can be trained on vast repositories of drug molecules, protein structures, disease pathways, and genetic data, enabling them to learn the intricacies of the human body at a level previously unattainable.

One of the most significant ways generative AI is accelerating drug discovery is by **designing novel drug candidates** . Imagine a machine that can conjure up entirely new molecules with desired therapeutic properties, not limited by existing libraries of compounds. This is the potential of generative models, which can create virtual drug molecules with specific targets in mind, bypassing the limitations of traditional screening methods. These models can explore vast chemical spaces, generating millions of potential drug candidates that align with desired properties, such as high potency, target specificity, and good pharmacokinetic profiles. This opens up an entirely new horizon in drug discovery, allowing researchers to venture beyond traditional approaches and explore uncharted territories in the search for effective treatments.

For instance, in the realm of cancer research, generative AI has shown remarkable promise in designing potent anti- cancer agents. By learning the intricate mechanisms of cancer cell growth and proliferation, AI models can generate molecules that specifically target cancerous cells while minimizing harm to healthy tissues. This approach has yielded promising candidates for targeting specific mutations in cancer cells, paving the way for personalized cancer therapies tailored to the genetic makeup of individual patients.

AI-driven virtual screening is another powerful tool for drug discovery, where generative AI can analyze vast libraries of existing compounds to identify those with high potential for therapeutic activity. Imagine a digital sieve that can sort through millions of molecules in a fraction of the time it would take a human researcher, pinpointing those with the best chance of becoming effective drugs. This is the power of AI-powered virtual screening, which can significantly reduce the time and cost associated with traditional screening methods.

Furthermore, generative AI can **predict the properties of drug candidates** , providing crucial insights into their potential efficacy, safety, and manufacturability. By simulating how molecules interact with biological targets, these models can predict drug efficacy and identify potential side effects before they are tested in humans. This predictive power can significantly reduce the risk and cost associated with drug development, allowing researchers to prioritize candidates with the highest potential for success.

For example, in the case of Alzheimer's disease, generative AI has been used to design novel inhibitors of amyloid-beta aggregation, a key pathological feature of the disease. AI models have been trained on vast databases of chemical compounds and their interactions with amyloid-beta, enabling them to generate potential inhibitors with high affinity and specificity for the target. This has led to the identification of promising drug candidates that are currently being evaluated in preclinical studies. However, the potential of generative AI in drug discovery extends beyond designing and screening molecules. It can also revolutionize **clinical trial design and execution** . By analyzing patient data and identifying key subgroups, AI models can help researchers recruit the most relevant participants for clinical trials, ensuring a higher probability of success and accelerating the path to bringing new treatments to market. One of the most compelling applications of generative AI in clinical trial design is its ability to predict patient response to treatment. By analyzing patient data, including genetic profiles, medical history, and lifestyle factors, AI models can identify patients who are likely to respond positively to a specific drug, allowing researchers to focus on these individuals in clinical trials. This targeted approach can lead to faster and more efficient trial execution, ultimately speeding up the development of new therapies.

But the impact of generative AI extends beyond the individual level; it has the potential to reshape the entire drug discovery landscape. By fostering collaboration between researchers and AI systems, it can create a more efficient and productive research environment. This collaboration can lead to breakthroughs in areas where traditional methods have struggled, accelerating the discovery of life-saving treatments for a wide range of diseases. The transformative power of generative AI in drug discovery is already being witnessed in real-world applications. Companies like Atomwise and Insilico Medicine are leveraging AI models to design novel drug candidates and accelerate the discovery process. These companies are working with pharmaceutical partners to develop AI- powered drug discovery platforms that are transforming the way new therapies are developed.

The future of drug discovery is undoubtedly intertwined with generative

AI. By unlocking new possibilities for drug design, virtual screening, clinical trial optimization, and collaborative research, AI is poised to accelerate the development of effective treatments for a wide range of diseases. The promise of AI-driven drug discovery is not just faster and more efficient drug development; it is the potential to unlock new frontiers in medicine and alleviate human suffering on a global scale.

This is a journey we are just beginning to embark on, and as we explore the uncharted territories of AI-powered drug discovery, we are witnessing a revolution in how we approach healthcare innovation. This revolution is not only about leveraging the power of machines but also about harnessing the human spirit of creativity, collaboration, and compassion to address some of the world's most pressing health challenges. The future of healthcare is not just about treating disease; it is about building a healthier future for all.

Personalized Medicine and AI

Imagine a world where medical treatments are no longer one-size-fits-all. A world where doctors can tailor treatments to individual patients, taking into account their unique genetic makeup, lifestyle, and medical history. This isn't just science fiction; it's the promise of personalized medicine, and AI is playing a pivotal role in making this vision a reality. At the heart of personalized medicine lies the idea that each individual's response to a treatment can vary significantly. This variation can be attributed to a complex interplay of factors, including genetics, lifestyle, and environmental exposures. Traditionally, healthcare has relied on a "one- size-fits-all" approach, often leading to less than optimal outcomes for some patients. Generative AI is transforming this landscape by offering powerful tools for understanding and predicting patient- specific responses. This is achieved through a variety of innovative applications, including:

AI-powered Drug Discovery and Development:

One of the most promising areas of AI application in personalized medicine is drug discovery. Traditionally, the process of discovering and developing new drugs is lengthy, expensive, and often unsuccessful. AI algorithms are now being used to accelerate this process by identifying potential drug targets, simulating drug interactions, and designing new drug candidates with greater precision. For example, generative models can analyze vast datasets of genetic information, disease characteristics, and existing drug databases to identify potential targets for drug development. These models can also be used to predict how a drug will interact with a patient's body based on their unique genetic profile, ultimately leading to more effective and targeted treatments.

Personalized Treatment Planning:

Generative AI is also playing a crucial role in developing personalized treatment plans. By analyzing patient data, including medical history, imaging scans, and genetic information, AI algorithms can identify the most effective treatments for individual patients.

For instance, in cancer treatment, AI models can analyze tumor biopsies and genetic information to predict the likelihood of response to different chemotherapy regimens. This allows oncologists to tailor treatment plans to each patient's specific needs, increasing the chances of successful therapy and reducing the risk of adverse side effects.

Precision Diagnostics and Risk Prediction:

Generative AI is not only revolutionizing treatment planning but also enhancing the accuracy and efficiency of diagnostics. AI algorithms can analyze medical images, such as X-rays, CT scans, and MRIs, to identify subtle patterns and abnormalities that may be missed by human eyes. This can lead to earlier diagnosis of diseases, allowing for more timely and effective interventions. Moreover, AI can be used to predict the risk of developing certain diseases based on a patient's genetic profile, lifestyle factors, and environmental exposures. This information empowers individuals to take proactive steps to manage their health, potentially preventing the onset of disease.

AI-powered Clinical Decision Support:

Generative AI is becoming an indispensable tool for clinical decision support. AI algorithms can analyze patient data, medical literature, and clinical guidelines to provide physicians with evidence-based recommendations for diagnosis and treatment. This can help doctors make more informed decisions by reducing diagnostic errors, optimizing treatment choices, and ensuring that patients receive the most appropriate care based on their individual needs.

Patient Engagement and Education:

Generative AI can also play a role in patient engagement and education. AI-powered chatbots and virtual assistants can provide patients with personalized information about their health conditions, medications, and treatment options. This allows patients to take a more active role in managing their health and making informed decisions about their care.

Beyond the Hype: Addressing Challenges and Ethical Concerns:

While the potential of AI in personalized medicine is undeniable, it's essential to acknowledge the challenges and ethical considerations that accompany this transformative technology.

Data Privacy and Security:

One of the key challenges is ensuring the privacy and security of patient data. As AI algorithms rely on vast amounts of personal health information, it's critical to establish robust data security measures and adhere to strict privacy regulations.

Algorithmic Bias:

Another concern is algorithmic bias. If AI models are trained on biased data, they may perpetuate existing inequalities in healthcare. It's crucial to address potential biases in training data and to develop algorithms that are fair and equitable.

Transparency and Explainability:

Transparency and explainability are also essential considerations. AI algorithms can be complex, making it difficult to understand their decision-making processes. For clinical applications, it's important to develop AI models that are interpretable and transparent, allowing clinicians to understand the rationale behind the recommendations provided by the algorithm.

Access and Equity:

Finally, it's important to ensure that the benefits of AI in personalized medicine are accessible to all patients, regardless of their socioeconomic status, geographical location, or insurance coverage.

The Future of Personalized Medicine:

Despite these challenges, the future of personalized medicine holds tremendous promise. With continued research and development, generative AI is poised to revolutionize healthcare, leading to more effective treatments, earlier diagnoses, and better patient outcomes. Imagine a world where every individual receives personalized care based on their unique needs, maximizing the benefits of treatment and minimizing the risk of adverse effects. This is the future that personalized medicine, powered by AI, is striving to achieve. The journey towards this future will require a collaborative effort among researchers, clinicians, ethicists, and policymakers. By working together, we can harness the power of AI to create a healthcare system that is truly patient-centered, equitable, and effective.

As we navigate this evolving landscape, it's essential to keep the human element at the heart of our approach. While AI can provide powerful insights and tools, it's ultimately the human touch that will guide us towards a future where healthcare is truly personalized and transformative.

AI and Genomics

The marriage of AI and genomics has ushered in a new era of personalized healthcare, transforming how we understand and treat diseases. AI's prowess in analyzing massive datasets, identifying patterns, and making predictions has revolutionized genomic research, unlocking a wealth of insights that were previously unimaginable. Imagine a future where your doctor can analyze your entire genome, identify specific genetic variations, and predict your susceptibility to certain diseases. With this information, personalized treatment plans can be tailored to your unique genetic makeup, maximizing efficacy and minimizing side effects. This is the promise of precision medicine, and AI is playing a pivotal role in making this vision a reality.

One of the most remarkable applications of AI in genomics is in drug discovery. Traditionally, drug development has been a long, arduous, and expensive process. But AI is changing the game by accelerating this process significantly. By analyzing vast libraries of chemical compounds and genetic data, AI algorithms can predict which molecules are likely to bind to specific target proteins involved in disease pathways. This enables researchers to prioritize promising drug candidates, significantly reducing the time and resources needed for preclinical testing.

Furthermore, AI is enabling the development of novel therapies. AI-powered tools are being used to design personalized gene therapies, which involve altering specific genes to treat genetic disorders. This approach holds immense promise for treating diseases like cystic fibrosis, Huntington's disease, and sickle cell anemia. Beyond drug discovery, AI is revolutionizing our understanding of disease mechanisms and evolution. By analyzing large-scale genomic datasets, AI can identify genetic mutations that are associated with specific diseases. This information can help researchers understand the underlying causes of diseases and develop targeted therapies that address those causes.

AI-powered tools are also empowering researchers to decipher the complex interplay between genetics, environment, and disease. By analyzing the genetic makeup of individuals alongside their lifestyle factors, AI can identify specific environmental exposures that may trigger or exacerbate certain diseases. This knowledge is crucial for developing preventive strategies and personalized healthcare recommendations.

The impact of AI on genomics extends beyond research to the clinical setting, where it is revolutionizing personalized medicine. AI-driven diagnostic tools can analyze patient data, including genetic information, medical history, and symptoms, to make more accurate and timely diagnoses. This allows for earlier intervention and potentially better outcomes for patients. Another crucial aspect of personalized medicine is the development of AI-powered tools for patient risk assessment. By analyzing an individual's genetic profile and medical history, these tools can predict their likelihood of developing specific diseases. This enables doctors to provide personalized prevention strategies and early interventions, reducing the risk of disease progression.

The integration of AI into genomic research and personalized healthcare is not without challenges. One major concern is the ethical implications of using AI to analyze sensitive genetic data. Ensuring data privacy and security is paramount, and robust mechanisms must be in place to protect patient confidentiality. Another challenge is the need for transparency and explainability in AI models used for medical decision- making. While AI can make complex predictions, it is essential to understand the reasoning behind those predictions. Developing AI models that are transparent and explainable will increase trust in AI-powered healthcare solutions. Despite these challenges, the future of AI in genomics is incredibly promising. As AI technologies continue to evolve, we can expect even more breakthroughs in personalized healthcare. This includes the development of more accurate and efficient diagnostic tools, personalized treatment plans tailored to individual genetic makeup, and the discovery of new therapies for previously untreatable diseases. However, it is crucial to approach the development and implementation of AI in genomics with a focus on ethical considerations and responsible innovation. Ensuring that AI is used to benefit humanity while safeguarding patient privacy and promoting equitable access to personalized healthcare is a critical responsibility for researchers, developers, and policymakers alike.

The journey of AI in genomics is just beginning, but the potential impact on human health is vast. As we harness the power of AI to unlock the mysteries of the genome, we are poised to enter a new era of personalized medicine that holds the promise of better health and well-being for all.

Advanced Imaging and Diagnostics

The realm of medical imaging has undergone a remarkable transformation with the advent of generative AI. Traditionally, medical imaging relied heavily on human interpretation of scans like X-rays, MRIs, and CT scans. However, AI is now revolutionizing this process, enhancing both diagnostic accuracy and efficiency. Imagine a world where AI can not only detect subtle anomalies in images but also predict the likelihood of certain diseases, leading to earlier interventions and potentially life-saving outcomes. This is the reality that generative AI is bringing to the forefront. One of the most compelling applications of generative AI in medical imaging is in the identification of subtle anomalies. Human radiologists, despite their expertise, can occasionally miss these subtle signs, leading to delayed diagnosis or misdiagnosis. Generative AI algorithms can learn from vast amounts of data, including normal and abnormal scans, and identify patterns that may not be readily apparent to the human eye. These algorithms can analyze images for minute changes in texture, shape, or density, providing insights that might otherwise go unnoticed. This enhanced sensitivity is particularly important in diagnosing conditions like early- stage cancers, where timely detection can significantly improve treatment outcomes. Beyond detection, generative AI is also playing a crucial role in the segmentation and analysis of medical images. Segmentation involves separating different structures within an image, like organs or tumors, from the surrounding tissue. This process, traditionally done manually by radiologists, is tedious and time-consuming. Generative AI models can automate this process, significantly reducing the workload on healthcare professionals and allowing them to focus on more complex tasks. AI-driven segmentation algorithms can identify and delineate structures with remarkable precision, enabling more accurate volume calculations and improving the effectiveness of treatment planning. The impact of generative AI extends to various medical imaging modalities. For instance, in mammography, AI algorithms are being trained to differentiate between benign and malignant breast lesions, helping radiologists identify cancer at earlier stages. In radiology, AI models can aid in detecting pneumonia and other lung diseases by analyzing chest X-rays, contributing to faster and more accurate diagnoses. Generative AI is also transforming the field of ophthalmology, with AI-powered algorithms assisting in the detection of diabetic retinopathy and other eye diseases. This is especially crucial in settings where access to ophthalmologists is limited.

Generative AI's role in medical imaging goes beyond just diagnosis. It is also being used to generate synthetic images that can be used for training new AI models or augmenting existing datasets. This is particularly important when dealing with rare diseases, where obtaining enough real patient data can be challenging. AI-generated images can create a more diverse and robust dataset, allowing AI models to be trained more effectively and achieve better performance in real-world scenarios.

The promise of generative AI in medical imaging is immense. It has the potential to improve diagnostic accuracy, reduce the workload on healthcare professionals, and ultimately, enhance patient care. While the technology is still in its early stages, ongoing research and development are rapidly pushing the boundaries of what's possible. Generative AI is already transforming the way we approach medical imaging, and its future impact on patient health is likely to be profound. However, it's crucial to acknowledge the ethical considerations surrounding the use of AI in healthcare. Ensuring data privacy and security is paramount. Additionally, transparency and explainability are essential for building trust in AI-powered diagnostics. The field is working towards developing ethical guidelines and ensuring that AI is used responsibly and for the benefit of patients. The future of medical imaging with generative AI is bright. As algorithms continue to improve and datasets grow, AI's role in diagnosis and treatment will become even more central. We are entering an era where AI and human expertise will collaborate to provide more accurate, efficient, and personalized healthcare for all.

AIs Role in Public Health

The ability of AI to analyze massive datasets and identify patterns has opened up incredible opportunities in the field of public health. By leveraging these capabilities, AI can play a pivotal role in managing and predicting health trends, ultimately improving public health outcomes. One of the most promising applications of AI in public health is in disease surveillance and outbreak prediction. Traditional methods of disease surveillance often rely on manual data collection and analysis, which can be time- consuming and prone to errors. AI algorithms, on the other hand, can process vast amounts of data from various sources, including electronic health records, social media, and environmental sensors, in real-time. This allows for the identification of emerging health threats and the rapid deployment of interventions. For example, researchers have developed AI models that can detect influenza outbreaks weeks earlier than traditional methods, allowing for timely public health interventions and potentially saving lives. By analyzing data on Google search queries, social media posts, and weather patterns, these models can predict the spread of influenza and other infectious diseases with high accuracy.

AI can also be used to analyze and predict the spread of chronic diseases, such as heart disease, diabetes, and cancer. By identifying risk factors and patterns associated with these diseases, AI models can help public health officials develop targeted interventions and preventive strategies. For instance, AI-powered tools can identify individuals at high risk of developing heart disease based on factors like age, genetics, lifestyle, and environmental exposures. This information can be used to tailor interventions and provide personalized recommendations to reduce their risk. Moreover, AI is revolutionizing the field of drug discovery and development. By analyzing vast databases of molecular structures and biological pathways, AI can identify potential drug candidates and accelerate the development of new treatments for various diseases. This can significantly reduce the time and cost associated with traditional drug discovery processes, ultimately leading to faster access to life-saving medications.

AI is also transforming public health communication and education. By leveraging natural language processing and machine learning, AI-powered chatbots and virtual assistants can provide personalized health information and support to individuals. These tools can answer questions, provide health recommendations, and track progress on health goals. This personalized approach can significantly improve health literacy and encourage individuals to take charge of their health. For instance, AI-powered chatbots can be used to provide information on vaccine safety and efficacy, debunking myths and addressing public concerns. They can also be deployed in healthcare settings to provide patients with personalized instructions and support during their recovery process.

The use of AI in public health presents several ethical considerations that need to be carefully addressed. Data privacy and security are paramount, as AI algorithms rely on large amounts of sensitive personal information. It is essential to ensure that data is anonymized and used responsibly, respecting individual privacy and safeguarding against potential misuse.

Another ethical concern is the potential for algorithmic bias. AI models can be trained on biased datasets, leading to discriminatory outcomes. For instance, if a disease prediction model is trained on data from predominantly white populations, it may not accurately predict health outcomes for individuals from other racial or ethnic groups. It is crucial to develop and deploy AI models that are fair, unbiased, and equitable.

Transparency and accountability are also essential when using AI in public health. It is important to understand how AI algorithms make decisions and to hold developers accountable for any biases or errors. Transparent and explainable AI models are crucial to ensure public trust and acceptance. To fully realize the potential of AI in public health, it is essential to foster collaborations between researchers, public health officials, policymakers, and industry experts. These partnerships can accelerate the development and deployment of AI-powered solutions, addressing challenges and ensuring ethical and responsible use. The future of public health is intertwined with the advancements in artificial intelligence. AI has the potential to transform how we prevent, manage, and treat diseases, improving health outcomes for individuals and communities worldwide. However, it is crucial to approach AI with caution, addressing ethical considerations and ensuring responsible development and deployment. By harnessing the power of AI responsibly, we can unlock its potential to create a healthier and more equitable world for all.

Chapter 5: Preparing for the Future with Generative AI

The future is being rewritten with generative AI, and it's imperative that we prepare ourselves for this transformative shift. This preparation begins with education and fostering AI literacy across all walks of life. Imagine a world where everyone, from students to business leaders, possesses a fundamental understanding of how AI works, its capabilities, and its potential impacts. This vision isn't a utopian dream; it's a tangible reality within our grasp. AI education isn't just about technical expertise; it's about cultivating a mindset that embraces the possibilities while navigating the complexities of this evolving technology. It's about demystifying AI, making it accessible, and empowering people to engage in informed discussions about its ethical implications, social impact, and future applications.

Here's why AI education is paramount:

Empowering the Workforce: In a future where AI is deeply integrated into various industries, a workforce equipped with AI literacy will be better prepared to thrive. They can leverage AI tools effectively, understand its limitations, and collaborate with AI systems to solve problems and drive innovation.

Driving Innovation and Entrepreneurship: AI education fosters an environment where individuals can conceive and implement new ideas powered by AI. It empowers individuals to be not just users of AI but also creators of AI-driven solutions, leading to a flourishing ecosystem of entrepreneurship and technological advancements.

Addressing Ethical Challenges: As AI becomes more pervasive, ethical considerations will become increasingly important. An educated populace can engage in informed discussions about the responsible development and deployment of AI, ensuring its benefits are widely shared while minimizing potential risks.

Promoting Responsible Use: AI literacy is crucial for navigating the complex landscape of AI applications. It empowers individuals to be discerning consumers, understand the biases inherent in AI systems, and make informed decisions about how they interact with AI-powered tools.

Here are some concrete steps we can take to foster AI education and literacy:

Integrating AI into Curricula: Schools and universities can introduce AI concepts into existing curricula across various disciplines, from computer science to arts and humanities. This approach helps students understand AI's potential applications in their fields and prepares them for the future workforce.

Developing Accessible Learning Resources:

Creating engaging and readily available online courses, tutorials, and educational materials makes AI education accessible to a broader audience. This includes catering to different learning styles and ensuring that resources are available in multiple languages.

Public Awareness Campaigns: Public outreach initiatives can help demystify AI, dispel myths, and build public trust in the technology. These campaigns can be targeted to various audiences, from children to professionals, using diverse mediums like documentaries, social media, and public events.

Encouraging Lifelong Learning: AI is evolving rapidly, so continuous learning is crucial for staying up-to-date. Encourage professionals to participate in workshops, conferences, and online courses to deepen their understanding of AI and its applications in their respective fields.

Building AI-Savvy Communities: Establishing online and offline communities dedicated to AI education and discussion fosters collaboration, knowledge sharing, and the development of best practices. These communities can also serve as platforms for tackling ethical concerns and navigating the evolving landscape of AI.

The Importance of AI Literacy for All

AI literacy isn't solely for tech professionals; it's essential for everyone. It empowers citizens to participate in informed discussions about the societal impact of AI, advocate for responsible development, and leverage AI for their own benefit. In the future, AI literacy will be as important as reading and writing, enabling individuals to fully participate in a world increasingly shaped by artificial intelligence.

Example: From Art to Healthcare

Imagine a young student fascinated by the intricate details of a painting. With AI education, they can learn how AI algorithms are used to generate art, understand the creative process behind AI-powered art, and perhaps even try their hand at creating their own AI-generated art pieces. This experience not only sparks creativity but also builds a foundation for understanding the potential of AI in various fields. Similarly, a healthcare professional can use their AI literacy to learn about the use of AI in personalized medicine. They can understand how AI algorithms analyze patient data to create tailored treatment plans and explore the potential of AI-powered diagnostics in improving patient outcomes. This knowledge allows them to collaborate effectively with AI systems and make informed decisions about incorporating AI into their practice.

The Power of Collaborative Innovation

AI education fosters a culture of collaboration between humans and AI. By understanding the strengths and limitations of both, individuals can leverage AI as a powerful tool for problem-solving and innovation.

Example: The Power of AI-Human Collaboration

Imagine a team of architects working on a sustainable city design project. They can utilize AI to analyze massive amounts of data, generate various design options, and simulate the environmental impact of each design. The team can then collaborate with the AI to refine these options, ensuring the designs meet their goals for sustainability, livability, and aesthetics. This symbiotic partnership harnesses the power of AI while retaining human creativity and judgment.

AI: A Tool for Empowering Every Individual

AI literacy isn't just about preparing for the future; it's about empowering individuals to shape that future. By fostering a society where everyone possesses a basic understanding of AI, we can ensure its responsible development and unleash its full potential for the benefit of all.

The journey to an AI-literate society requires a collective effort. Governments, educational institutions, technology companies, and individuals all have a role to play in making AI education accessible and empowering people to embrace the opportunities and navigate the challenges of this transformative era.

Looking Ahead: AI Literacy for a Better Future

The future of AI is bright, but it depends on our ability to prepare for the changes it will bring. By prioritizing AI education and literacy, we can create a future where AI is a powerful force for good, enhancing creativity, solving complex problems, and ultimately improving the lives of everyone.

Collaborating with AI

The era of generative AI presents a unique opportunity for humans and AI to collaborate in unprecedented ways. This partnership is not about replacing human skills but rather about augmenting them, creating a synergy that unlocks new realms of innovation and efficiency.

A Symphony of Creativity

Imagine a composer struggling to find the perfect melody for their symphony. By collaborating with an AI-powered music composition tool, they can input their initial ideas, explore variations, and receive AI-generated suggestions that inspire new directions. The AI, trained on vast musical datasets, can analyze patterns, suggest harmonies, and even experiment with novel soundscapes, pushing the composer's creative boundaries. This partnership allows the composer to focus on the emotional core of their music while the AI handles the technicalities, resulting in a richer, more nuanced composition.

Revolutionizing Problem-Solving

In the realm of scientific research, generative AI can act as a powerful research assistant. Imagine a scientist investigating a complex disease. They can use AI to analyze vast medical datasets, identify potential drug targets, and even design new drug molecules. The AI can sift through millions of data points, finding correlations and patterns that humans might miss. This collaboration allows scientists to accelerate research, develop personalized treatments, and ultimately save lives.

Augmenting Expertise

Generative AI can also be used to augment human expertise in fields like writing, design, and coding. A writer working on a historical novel could use AI to generate realistic dialogue based on historical records or to research specific events. A designer could use AI to create prototypes and explore different design concepts, speeding up the design process. A software developer could use AI to generate code snippets or to identify potential bugs in their code. In each of these cases, AI acts as a tool that empowers humans to achieve greater efficiency and creativity.

The Key to Effective Collaboration

The key to a successful human-AI partnership lies in understanding the strengths and limitations of each. Humans bring creativity, intuition, and the ability to make subjective judgments. AI excels at processing massive amounts of data, identifying patterns, and executing tasks with speed and precision. By recognizing these distinct strengths, we can create a harmonious collaboration where humans and AI complement each other.

The Future of Collaboration

As generative AI continues to evolve, the possibilities for collaboration will only expand. Imagine a future where AI can generate personalized learning experiences for students, create customized marketing campaigns for businesses, or even design personalized medical treatments for patients. The potential is vast, and the future of human-AI collaboration is truly exciting.

The Importance of Human Control and Oversight

While the benefits of generative AI are undeniable, it's crucial to remember that AI is a tool, and like any tool, it can be used for good or bad. The ethical implications of AI are complex and deserve careful consideration. It's essential to maintain human control over AI systems, ensure transparency in their decision-making, and hold them accountable for their actions.

Empowering Humans in the AI Age

Generative AI is not meant to replace humans. It's meant to empower them. By understanding how to effectively collaborate with AI, we can harness its incredible potential to create a future where technology and human creativity work together to solve complex problems, push the boundaries of innovation, and ultimately create a better world for all.

Key Takeaways:

The future of generative AI lies in collaboration between humans and machines. Humans bring creativity, intuition, and subjective judgment, while AI excels at data analysis, pattern recognition, and speed. This partnership can revolutionize various fields, including creative arts, scientific research, business, and medicine. It's essential to maintain human control and oversight of AI systems to ensure ethical and responsible use. Generative AI has the potential to empower humans to achieve greater heights of creativity, innovation, and progress.

Moving Forward:

As we embrace the age of generative AI, it's crucial to invest in AI education and literacy to equip individuals with the skills and understanding necessary to navigate this evolving landscape. By embracing the potential of this powerful technology, we can create a future where human ingenuity and AI collaboration flourish. The future of innovation belongs to those who are ready to collaborate with the machines.

Innovative Leadership in the AI Era

The world is rapidly evolving, propelled by the transformative power of artificial intelligence (AI). This evolution is not merely technological; it's also a profound shift in how we work, think, and lead. In this age of generative AI, where machines can create and innovate alongside humans, leadership takes on a new dimension. To thrive in this dynamic landscape, leaders must embrace a set of qualities and approaches that are as innovative as the technologies they harness.

From Command and Control to Collaboration and Empowerment

The traditional leadership paradigm of command and control is becoming increasingly obsolete. AI is not a tool to be simply directed; it's a partner, a collaborator, and a source of boundless creativity. Leaders need to cultivate an environment where AI is embraced as an integral part of the team, where human ingenuity and AI's computational power work in tandem. This requires a shift in mindset. Instead of viewing AI as a threat to human jobs, leaders must recognize its potential to augment human capabilities, opening doors to new possibilities and unlocking greater efficiency. It's about fostering a spirit of collaboration, where humans and AI work together to achieve shared goals.

Embrace Continuous Learning and Adaptability

In the AI-driven world, the pace of change is relentless. The algorithms and models that power generative AI are constantly being refined, leading to new breakthroughs and applications. Leaders who want to stay ahead of the curve must adopt a mindset of continuous learning. This involves actively engaging with the latest developments in AI, understanding how these advancements can be applied to their specific fields, and encouraging a culture of ongoing exploration and adaptation within their organizations.

This adaptability extends beyond just technical skills. Leaders must be willing to adapt their strategies, organizational structures, and even their own roles in response to the changing landscape. The ability to anticipate and respond to the disruptive potential of AI is crucial for navigating this dynamic future.

Data Literacy and Ethical Stewardship

Generative AI thrives on data. It learns from vast datasets, extracting patterns and insights to generate new creations. Leaders must understand the importance of data in this context. This involves developing a strong understanding of data literacy, which encompasses not just the ability to collect and analyze data but also the ability to interpret its meaning and implications. Beyond technical prowess, leaders must also embrace the ethical responsibilities that come with AI. Generative models can be powerful tools, but they also carry the potential for misuse. Leaders must actively guide the development and deployment of AI in an ethical and responsible manner, ensuring fairness, transparency, and accountability.

Building a Culture of Creativity and Innovation

Generative AI is not just about automation; it's about unlocking new frontiers of creativity and innovation. Leaders must foster a culture that encourages experimentation, risk-taking, and a willingness to embrace the unexpected. This means creating an environment where employees feel empowered to explore new ideas, fail fast, and learn from their mistakes. It's about providing the resources and support needed for teams to experiment with generative AI, explore its potential, and translate those insights into tangible solutions.

Leading with Empathy and Emotional Intelligence

The transition to an AI-driven world is not without its challenges. As AI plays a more prominent role in various industries, questions about the future of work and the impact on human roles will inevitably arise. Leaders must lead with empathy and emotional intelligence, understanding the concerns and anxieties of their teams. They must be sensitive to the human element, providing clear communication, support, and training to navigate this evolving landscape.

Embracing the Transformative Power of AI

The rise of generative AI is a remarkable chapter in human history. It's a testament to our ingenuity and our ability to create tools that can push the boundaries of what's possible. Leaders who embrace this transformative power, who understand the qualities and approaches needed to navigate this dynamic future, will be the ones who shape the world of tomorrow. This is not just about technology; it's about harnessing the power of AI to create a more equitable, sustainable, and innovative future for all. By fostering collaboration, adaptability, ethical stewardship, and a spirit of creative exploration, leaders can unlock the full potential of generative AI and guide humanity towards a brighter future.

AI and Societal Impact

The widespread adoption of AI, especially generative models, is poised to reshape our society in profound ways, both positive and challenging. Understanding these potential impacts is crucial for navigating the future and ensuring that AI benefits humanity.

Economic Disruption and Job Market Transformations

One of the most immediate impacts of generative AI will be on the job market. Automation powered by AI is already replacing certain tasks in various industries, and this trend is likely to accelerate as generative models become more sophisticated. While some jobs may be eliminated, others will emerge as new roles emerge in AI development, maintenance, and ethical oversight. The transition will require significant workforce training and retraining programs to equip individuals with the skills necessary to adapt to the changing labor landscape. The societal challenge lies in ensuring equitable access to such training and preventing widespread unemployment as AI reshapes the economy.

Augmenting Human Capabilities

Generative AI has the potential to augment human capabilities in diverse fields. In healthcare, AI can analyze massive datasets to identify patterns and develop personalized treatment plans, leading to improved patient outcomes. In education, AI can personalize learning experiences for each student, adapting to their individual needs and pace. In scientific research, AI can assist in complex simulations and data analysis, accelerating the pace of discovery. The key lies in harnessing the power of AI to enhance human capabilities, not to replace human judgment and creativity.

Ethical Concerns and Bias Mitigation

As generative AI becomes more prevalent, ethical concerns arise. One major challenge is the potential for bias in AI systems. AI models are trained on data, and if that data reflects societal biases, the AI system will inherit those biases, potentially leading to discriminatory outcomes. Ensuring fairness and equity in AI development and deployment requires rigorous efforts to mitigate bias. This involves using diverse datasets, developing methods for identifying and mitigating bias, and ensuring transparency in AI decision-making processes.

Privacy and Data Security

Generative AI relies heavily on vast datasets, raising concerns about privacy and data security. The collection, storage, and use of personal data must be governed by stringent regulations and ethical principles. Ensuring data privacy and security is paramount, and individuals should have control over their personal data. Secure data management practices, privacy-preserving technologies, and robust security measures are crucial for building trust in AI systems.

Social and Cultural Impacts

Generative AI is also poised to have a profound impact on social and cultural spheres. AI-generated art, music, and literature will challenge traditional notions of creativity and authorship. The rise of deepfakes and AI-generated content could lead to challenges in discerning truth from falsehood. Societal conversations about authenticity, ownership, and the nature of creativity will be crucial. Ethical guidelines and regulations are needed to address the challenges of misinformation and manipulation arising from AI-generated content.

The Future of Work and Society

The future of work and society in an AI-driven world depends on how we shape the development and deployment of generative AI. Fostering collaboration between humans and AI will be essential. We need to develop AI systems that are reliable, transparent, and accountable, ensuring that they augment human capabilities and serve the common good. Investing in education and training, promoting ethical guidelines for AI development, and fostering open dialogues about the social and cultural impacts of AI are crucial for navigating the future with generative AI.

The Power of Generative AI: A Force for Good

Generative AI holds immense potential to drive progress across diverse domains. From healthcare and education to scientific research and art creation, AI can empower us to solve complex challenges and create a better future. However, realizing this potential requires careful consideration of the ethical and societal implications of AI. By embracing responsible innovation, ensuring equitable access to AI benefits, and fostering a future where humans and AI collaborate to solve global challenges, we can harness the transformative power of generative AI for the betterment of society. The journey into the AI-powered future is one that requires both technological advancements and societal adjustments. As generative AI continues to evolve, it is our responsibility to ensure its development and application align with human values, promoting a future where AI enhances our capabilities, strengthens our communities, and creates a more just and equitable world.

Charting the Path Forward

The future of AI and generative technologies holds immense promise, a landscape teeming with exciting possibilities and transformative potential. As we stand on the cusp of a new era where machines possess the power to create, innovate, and shape our world, it's imperative to navigate this future with a clear understanding of its implications and a proactive approach to its development.

Embracing the AI Revolution:

Imagine a world where AI-powered tools become indispensable partners in our creative endeavors, amplifying our human ingenuity and pushing the boundaries of imagination. This is not a distant sci-fi fantasy but a reality rapidly approaching. The advent of generative AI has already begun to redefine industries, from art and music to medicine and engineering. Its ability to learn from data, generate new ideas, and solve complex problems has unlocked a plethora of possibilities that were once considered exclusive to the human mind.

AI Literacy: A Prerequisite for the Future:

Preparing for this future necessitates a fundamental shift in our approach to education. AI literacy, the ability to understand and interact effectively with AI technologies, will become a crucial skillset for individuals across all disciplines. We need to equip ourselves and future generations with the knowledge and tools to engage meaningfully with AI, to understand its strengths and limitations, and to harness its power for good.

Human-AI Collaboration: A New Era of Partnership:

Instead of viewing AI as a threat to our own creativity, we should embrace it as a collaborator, a powerful tool that can augment our abilities and accelerate innovation. The future lies in fostering a symbiotic relationship between humans and AI, where each brings its unique strengths to the table. Humans, with their intuition, empathy, and critical thinking skills, can guide AI's creative endeavors, ensuring that its outputs are aligned with ethical values and human needs.

Leadership in the AI Era:

Leaders in this new era of AI will need to possess a distinct set of skills and perspectives. They will be those who embrace innovation, champion collaboration between humans and AI, and champion the responsible development and deployment of these powerful technologies. Ethical considerations must be at the forefront of every decision, ensuring that AI is used for the betterment of humanity and not for its detriment.

Navigating the Societal Impact of AI:

As AI technologies become increasingly integrated into our lives, we must consider their wider societal implications. Will AI exacerbate existing inequalities or create new ones? Will it lead to the displacement of jobs or the creation of new opportunities? These are crucial questions that demand careful consideration and proactive policy-making to ensure that AI benefits all of society, not just a select few.

A Path Forward: Ethical, Inclusive, and Sustainable AI:

The future of AI is not predetermined; it's a path we can actively shape. By prioritizing ethical development, promoting inclusivity in AI research and deployment, and fostering sustainable practices in its development, we can ensure that AI becomes a force for good, driving progress, fostering creativity, and improving the lives of all.

Investing in AI Research and Development:

Continued investment in AI research is vital to unlock its full potential and address its challenges. We need to support the development of new AI algorithms, foster innovation in AI hardware, and encourage interdisciplinary research that explores the ethical, societal, and economic implications of AI.

Creating a Global AI Ecosystem:

Collaboration across borders and disciplines is crucial for unlocking the full potential of AI. International partnerships and shared research initiatives can accelerate the pace of progress, foster knowledge exchange, and ensure that the benefits of AI are distributed equitably across the globe.

Building a Future Where AI Empowers Humanity:

The future of AI is not merely about the advancement of technology but about the advancement of humanity. By embracing AI as a tool for creativity, collaboration, and progress, we can harness its transformative power to create a brighter, more sustainable, and more equitable future for all.

References

1. Bengio, Y., LeCun, Y., & Hinton, G. (2015). "Deep Learning." Nature, 521(7553), 436–444.
2. Brown, T., Mann, B., Ryder, N., et al. (2020). "Language Models are Few-Shot Learners." Advances in Neural Information Processing Systems (NeurIPS).
3. Goodfellow, I., Bengio, Y., & Courville, A. (2016). Deep Learning. MIT Press.
4. Kingma, D. P., & Welling, M. (2014). "Auto-Encoding Variational Bayes." arXiv preprint arXiv:1312.6114.
5. Radford, A., Wu, J., Amodei, D., et al. (2019). "Better Language Models and Their Implications." OpenAI.
6. Vaswani, A., Shazeer, N., Parmar, N., et al. (2017). "Attention Is All You Need." Advances in Neural Information Processing Systems (NeurIPS).
7. Zhang, H., Goodfellow, I., Metaxas, D., & Odena, A. (2019). "Self-Attention Generative Adversarial Networks." Proceedings of the International Conference on Machine Learning (ICML).
8. OpenAI. (2023). "The Economics of Generative AI."
9. Marcus, G. & Davis, E. (2019). Rebooting AI: Building Artificial Intelligence We Can Trust. Pantheon.
10. Ng, A. (2021). "AI Transformation Playbook: How to Lead Your Company Into the AI Era."
11. Dosovitskiy, A., Beyer, L., Kolesnikov, A., et al. (2021). "An Image Is

Worth 16x16 Words: Transformers for Image Recognition at Scale." Proceedings of the International Conference on Learning Representations (ICLR).
12. Ethics in AI Research. (2022). "Responsible AI: Addressing Bias and Ensuring Fairness." Journal of Artificial Intelligence Research (JAIR), 72, 1–15.
13. Hao, K. (2020). "AI and Creativity: How Artificial Intelligence is Changing Art and Design." MIT Technology Review.
14. Floridi, L., & Cowls, J. (2019). "The Ethics of AI: A Framework for Discussion." Nature Machine Intelligence, 1(4), 261–263.
15. Sutton, R. S., & Barto, A. G. (2018). Reinforcement Learning: An Introduction. MIT Press.

www.ingramcontent.com/pod-product-compliance
Lightning Source LLC
Chambersburg PA
CBHW040318220526
45473CB00009B/2485